W9-BKP-820

BISHOP'S UNIVERSITY
LIBRARY
LENNOXVILLE

THE LONG TUNNEL

THE LONG

NEW YORK 1976 ***Atheneum***

TUNNEL:

A Coal Miner's Journal

MEADE ARBLE

Library of Congress Cataloging in Publication Data
Arble, Meade.
The long tunnel.

1. Coal-miners—Pennsylvania—Personal narratives.
2. Arble, Meade. I. Title.
HD8039.M62U6125 1976 331.7'62'330924 [B] 76-11526
ISBN 0-689-10738-2

The names of many of the people and places in this book have been changed.

Parts of this book have appeared, in different form, in the *New York Times Magazine* and *Redbook*.

Published simultaneously in Canada by
McClelland and Stewart Ltd.
Composition by Dix Typesetting Co. Inc., Syracuse, New York
Printed and bound by The Murray Printing Company,
Forge Village, Massachusetts
Designed by Kathleen Carey
First Edition

To Grada

Few men realize that their life, the very essence of their character, their capabilities and their audacities are only the expression of their belief in the safety of their surroundings.

JOSEPH CONRAD

PART 1

THE CLOUDS BROKE AND PATCHES OF LATE AFTERNOON sun appeared as we turned into the dark, mushy lane past the heavy stone pillars holding the gate. A humid August wind waved the tops of the tall pines that shadowed the road. Hadley, four, and Bo, two, were asleep in the back seat, wrapped in a sweaty heap beside Baron, the big black German shepherd. Our '64 Chevy Impala, creaking under everything we owned, scraped the high, wet grass sparkling in the sunlight.

Faced with no job, no money, and no place to live, my wife, Grada, and I and our two children had come to northern Pennsylvania. I planned to find work in the coal mines, and we were going to stay in a cabin built over fifty years ago by my grandfather. The cabin, isolated among eighty-two acres of pines, had been used by my grandfather to hobnob with the magnates of Andrew Carnegie's era.

There had been a tennis court and a swimming pool. A stream that meandered around the cabin was used for canoeing, and the spacious front lawn had been studded with croquet wickets. My grandfather had been a doctor in Greenridge, a mining and farming town of 1,100 Germans and Poles three miles away. He died when I was four, and I remember him only as a stern, goateed face in a sepia tint. Both of his sons were also doctors. When he died, my father took over his practice, and my uncle, the younger brother, moved away.

In the old days the rows of pines on each side of the lane had been trimmed up to six feet, providing a clear view for a hundred yards back over a carpet of soft needles. Now the forest was a jungle of brush and dying trees. The tennis court and swimming pool were gone and the shallow stream was gray from mine effluence. My parents had sold everything, the office and the house, and moved to Arizona in the late sixties in hopes that it might help my father's emphysema. When he died in 1972, my mother gave her half of the cabin to me.

On our way to the cabin from Grada's parents in Westwood, New Jersey, we had stopped at my uncle's for the keys. He and his family were very unhappy. We were upsetting their vacation plans, they said. There had been no prior arrangement.

I explained that we wanted to live in the cabin only until I went into the coal mines and earned enough money to be able to rent an apartment. After all, I said, one cold water tap and an outhouse in that swampy basin were not ideal for young children, especially the newborn that would arrive in September. After an hour of strained pleasantries they gave up a set of keys and we continued the trip that had begun six hours earlier in New Jersey.

After a few hundred yards the lane opened into a parking circle. We stopped, and while Grada mopped her face and finished the Thermos of cold lemonade, I went across the old, rotted bridge and the wide lawn to the cabin. The heavy cabin logs were decayed, but solid. Towering pines kept the two stories in constant shade. Wild bees that had swarmed under one of the eaves made passes while I worked the padlock open on the door.

Inside, late afternoon light cast squares of linty brightness over the handmade, split-log furniture. Because of the big pines, the air was always dank and pungently cool, no matter how hot outside. Small cones of sawdust dotted the floor under the beams. I went back to the kitchen at the far end and took the circuit breakers off the nails under the sink. When the circuit breakers were plugged into the fuse box, lights came on and the refrigerator hummed.

I took an enamel pitcher up the hill in back that led to the spring and filled the water line. There was substantial pressure, and cold, clear water gushed from the tap. The old copper sink leaked slightly, but it was no problem.

Back across the swaying bridge, I draped a baby over each shoulder. Grada, awkward and heavy in her eighth month, followed me to the cabin. I put the kids upstairs on one of the four double beds and then unloaded the car while Grada inspected the kitchen. After we were settled, we sat on the front porch in a couple of old rocking chairs and drank gin and tonics.

It was Saturday, the weekend of the annual Greenridge Church Festival. I was anxious to reacquaint myself with the town, and the festival would be full of union men who might be able to help me get a job underground. Grada, however, was not eager to be left alone. The nearest telephone was about a mile away at a farm where they kept mean dogs. I told her I had to go in. It was a chance to see a lot of influential friends in one place.

While we talked, the last pale rays of sun slanted through the trees. Grada's glossy chestnut hair was brushed out long and shining and she wore a brightly flowered dress that she had bought in Spain. Her shawl was clutched around her as she rocked. When Grada is angry she won't look at me. She is half Spanish and half Dutch, and she does not cry easily. Her large hazel eyes become bright and the lids tremble, and her face loses color except for bright spots on her cheeks, but she usually keeps control.

"You'll be out all night," she said. "I know it. You'll get drunk with your old cronies and spend the night telling those old stories."

We had visited Greenridge four years ago, before leaving for Spain, and that had happened. I swore to her that it wouldn't happen again. What kind of man would leave his wife and babies in the woods alone while he went off drinking and carousing? Besides, the dog would take care of them.

"Look at yourself," she continued. "We've been married six years, and here we are, broke, I'm having another baby in a month, and we're in a cabin waiting for you to go into the coal mines. . . ."

"What the hell do you want?" I exploded. "I didn't promise you anything when we were married. You obviously took the best deal you could get at the time, and now you're stuck with it. You think I *want* to go into the mines?"

She shook her head. "A doctor's son with four years at Penn State and four years at the University of Seville shouldn't *have* to go into the mines to support his family. You're thirty-four, Meade. Can't you see what's happening?"

Of course I could see it. The pattern was familiar. When I was twelve, I regularly stole the family car and went joy riding, and once I even stole the scoutmaster's car. By the time I was fifteen, my parents had been so confounded by my behavior that they sent me to Staunton Military Academy in Virginia to shape up. Besides, they said, I would meet good contacts and make lifelong friends. Although my classmates included later luminaries such as John Dean III, Barry Goldwater, Jr., and folk singer Phil Ochs, none of them became fast friends. I narrowly graduated and achieved only the distinction of having been one of the school's worst cadets.

My father was determined to salvage me, his only son. My sister, two years older, had always been near the top of her class and was everything they hoped for. After Staunton I joined the navy with the hope that they might rehabilitate me. Soon after boot camp, aboard a destroyer escort in Pearl Harbor, I was put over the side on a stage to chip paint. Chipping was normally done in large blocks, and the exposed metal was protected at the end of the day with a coat of dayglo orange paint called red lead. I spent the day in an artistic frenzy chipping a life-size nude girl, hair flying in the wind, and at the end of the day I red-leaded her. She could be seen all the way across the harbor, shimmering in the blazing sunset. She was seen, in fact, by the harbor commodore, who phoned the captain of our ship for an explanation. That incident cost me the first of a couple trips to the brig, where the marines were into their own type of post-military school hazing as described by James Jones in *From Here to Eternity.*

After two years of similar hi-jinks, the navy and I called it a draw. I left the service in 1960 with a general discharge in San Francisco, California. From there I drifted around for a few months, riding freights, sitting in small town jails on vagrancy and disorderly charges, and finally returned home.

My father sent me to Penn State. I was determined to shape up. I met Grada, played the drums in rock bands, drank a lot, and left with no degree after four years. My average was too low. From there, it was on to New York City, a string of jobs, more booze, until finally, at age thirty, I decided that my only future was in dropping out, going to Spain, and writing a best-selling novel. By that time Grada and I were married and had a little girl. We sold most of the furniture, withdrew our $2,000 savings, and prepared to leave.

Before going to Spain, we flew out to Phoenix to visit my parents. My father, sixty-four, was in the last stages of emphysema, the result of two packs of Camels a day over a lifetime. He had a hospital bed in the apartment, oxygen, and a Bennett machine to breathe for him four times day. Some mornings he couldn't get out of bed to wash his face. Every breath was painful.

On our last afternoon in Phoenix, he lay in bed, puffing after every sentence. The rest of the family were out by the pool. He and I knew we would not see each other again. During that long afternoon I collected the payoff for a lifetime of building a wall around my emotions. At Staunton, when I was fifteen, they had shaved my head for a minor infraction three days before Christmas vacation. They got a little whining out of me, but not much. Later, in the brig, the jarheads who spit in my face while I stood at attention stripped naked received no reaction at all. By the time I was sitting with my father for the last time, I was able to spend the entire afternoon talking about nothing closer to either of us than the weather.

Then it was off to Spain. I enrolled in a university to collect the G.I. Bill and explained to everyone that I was a writer. After four years I had produced two short stories.

Not much, I admitted, but I was always a slow writer, and these stories were models of tight, controlled prose. However, no one bought them. The magazines liked them, but said that there were no contiguous epiphanies. They were old-fashioned.

By the time the G.I. Bill gravy train ran out, we had accumulated a pile of bills. My résumé was laughable, even with the standard assortment of lies, distortions, and half-truths. I had mailed out over a hundred résumés upon our return and did not receive a single response. Grada had a degree in social work, but she was pregnant. We returned in the middle of the '74 recession. We either had to go on welfare, which was starvation on the installment plan, or I had to jettison my carefully nurtured lies and start battling for survival.

When we first returned we lived with Grada's parents in their tiny house in Westwood, New Jersey. Grada's father had left his home in northern Spain, a place of absolute poverty, when he was thirteen, and arrived in the United States with nothing. He became a barber, and ended up owning his own shop, a house, two cars, and sending his two daughters through college. He saw nothing strange at all in my going into the coal mines. He wished us luck and gave us a few tips about living off the land.

Grada finally agreed that I should go to the festival. We brought over everything from the car and stacked the boxes and suitcases in the living room. While she made supper, I brought in wood and cleaned up the fireplace, a massive stone structure that dominated the room.

After supper, we put the kids to bed and I drove to

Greenridge. In the last light of evening, lines of street lights connected the scattered houses surrounded by dark patches of trees and yellow fields of wheat and oats.

Prosperity was evident. Many of the clapboard houses were aluminum sided, others were freshly painted. I parked beside the fire hall and walked up the hill to the festival on the asphalt lot between the church and the parochial elementary school. The streets held a clean, washed odor from the afternoon rain. Under rows of maples and elms, children rode their bikes through the puddles.

At the festival, there were kiddie rides and chance booths. Food stalls sent steamy smells of hot sausage, proagies, peppers, kielbasa, chiliburgers, sauerkraut, and pizza swirling into the air. Around the corner of the school, shattering blasts of rock music split the night.

The core of the festival was inside a snow fence that enclosed the beer stand, the Chuck-a-Luck, and the crap table. Gnats attacked the strings of light bulbs overhead while women with sleeping children on their laps sat talking on rows of wooden benches under canvas. The men clustered together drinking beer and swapping stories. I bought a half-dozen beer tickets and moved into the sweaty group of madras bermudas and tank shirts. Everyone was swilling beer and firing the empties over the crowd into a dump truck, or dropping the can, stepping on it, and popping another.

Jeb, a big-muscled trucker with black, wavy hair and the long sideburns of an 1890's bartender, was telling a story to Mitch and Ann. Jeb and I were exactly the same age. Ann, a year younger, had been our schoolmate. Though she was now divorced with two kids, she still had the long blond hair, freckles, and bright smile that had made her a seminal influence during our adolescence. Mitch, in his late forties, had raised twelve kids while putting in almost thirty years underground.

We shook hands. Jeb was winding into a story about his army days in Germany. I asked Mitch if he could help me get into the mines. He knew a lot of people in the union and at Nova Coal Co., a major local mine-operating outfit.

He said he would be glad to help. I told him it had better be before September 22, because our new baby was due then and we didn't have the money to pay for its delivery. He asked why I was going into the mines, and I said that I needed the money. I was ready to unload the sad tale of my life, but he didn't pursue the question. Later I was to find that simply needing the money was explanation enough for almost everyone there.

". . . downtown one night with no place to stay, and I met up with this hooker," continued Jeb, standing with a beefy arm on Mitch's shoulder. Ann smiled. Next to us, on the other side of the fence, a cop forced a teenager to pour a full beer out on the ground. "It was a hundred marks, about twenty-five dollars, to stay all night and get a little pussy, have some drinks. So we go to her place and there's this goddam dog there, keeps growlin' at me."

Shouts and cheers erupted from the crap table. A tractor putted by, pulling a wagon of hay bales and kids. I asked Mitch which mine I might get into. He said, "They're all the same, except #32, and that's high coal. But you probably won't get in there. You'll be out with me at 18-D or 18-B, where it's low and wet."

The tractor moved slowly past and the kids yelled and

waved. The area inside the snow fence was jammed. Groups of teen-agers in threes and fours hid their beers next to their legs and glanced around at the cops from neighboring towns helping out the one from Greenridge. The kids weren't arrested for drinking, but the cops made them pour out the beer and leave the fenced section.

". . . the sonofabitch laid there and growled, just a little bastard. Anyhow, she went out for some booze, and I thought sure as shit she's gonna rob me or somethin', so I took my wallet and hid it behind the bureau. She comes back with the stuff and asks me for the money. I'm feelin' like an asshole 'cause it's hid behind the dresser . . ."

Mitch's wife appeared from the mass around us and took his arm. We talked about how things were in the area, and I made the usual mistake of complimenting her on what great shape she was in for a mother of twelve. "Everybody says that," she said, disgusted. "Why can't I just be attractive without counting the kids?"

". . . so I paid her and we're getting into it, see, and afterwards I go to sleep." Ann had been joined by Jeb's wife and two of his four kids. She recognized an old story and tried to leave. Backing up, she bumped into Zurko, lying on a bench with a beer on his chest. He sat up and we went through the handshaking and the brief exchanges of what had happened to each other over the past ten years or so. In the shadows from the naked bulbs overhead, Zurko's face, lined and seamed around the hollows of high cheekbones, looked like an Indian's. His nose, broken several times, was flat and askew. He had cut himself shaving and specks of blood mottled the side of his face. Zurko had been a football star in high school and had gone through a state teachers college on scholarship. He had been several years ahead of me in high school, but we had drunk together a lot in our early twenties and had good times. I was surprised to learn that he was in the mines with almost nine years underground. He said that when he graduated, miners were making more money than teachers and he hated teaching anyhow. He had gotten married and gone into the mines temporarily, he had thought, to stash away some money.

". . . and I wake up in the middle of the night," said Jeb, "and I figure I'll take another one, so I reach over and put my hand down there by her snatch. I hear this goddam growlin'! The dog's down there, sleepin' between her legs . . ."

Zurko, in a torn T-shirt, still appeared to be made of whale bone and piano wire. Muscles and tendons bulged, and veins lined his forearms. I remembered him slamming through defensive lines, nose bloodied, eyes flashing with malice.

". . . I woke her up and she says, 'Oh, he always sleeps there.' " Jeb leaned forward, taking Ann into his confidence. "I tell it now that the dog bit me, but I'm giving you the earlier version . . ."

Zurko and I moved along the snow fence to a small space between the fence and the school, and left. I told him I had to go back to the cabin; my wife was very pregnant and the nearest phone was a mile away.

We went across the road that separated the school from the town graveyard, dark and silent under the trees. My father's grave was there beside my grandfather's, but I wanted to wait for a quiet day to visit it. Zurko had parked his flatbed truck alongside the cemetery. We decided to have another quick beer. He pulled out his cooler and two lawn chairs and set them up on the bed. It was the way he took the family to the drive-in, he said; just backed the truck in beside the speakers and set up the chairs and cooler.

Monday, I went to Tippleside, eight miles from Greenridge, to the Nova Company office, and added my name to a long list. Tippleside was larger than Greenridge; it had about a half-dozen traffic lights to our one. Tippleside also had a movie theater, reduced to showing porno flicks, and a municipal swimming pool. Greenridge had no pool, no movies, no pool hall, no dance hall. Its only diversions were bars and a bowling alley in the fire club. The personnel man at the Nova office asked if I had any family or relatives that worked for Nova, and if I was a local resident. I said no relatives, but I had been born and raised in Greenridge. He wished me luck. Depending on how much help I got, he said, the waiting list was three months long.

Back at the cabin, we settled in to wait. The kids played around the woods and in the stream, as I had at their age. The trout were long gone and now the stream was gray, but Hadley and Bo still crouched in their tennis shoes in the cold water with their toy buckets, watching for minnows and crayfish.

By September the weather had turned cold. In the early mornings steam rose from the chamber pot when I emptied it in the woods. White mist hung over the creek. Since we had no phone, I had been calling the Nova office every few days during my trips to the post office. On September 7, my birthday, I called and was told that I would start on September 16. I would be required to attend a three-day safety school beginning on that date.

We celebrated with a bottle of champagne at the cabin. Grada, now extravagantly pregnant, was vastly relieved that my starting date gave us a week's grace before the baby was due.

On a sunny day, about a week before I was to begin work, we went to Tippleside to shop for miner's work clothes. In an army surplus store we picked up a canvas jacket, $15; canvas trousers, $15; two sets of thermal underwear, $20 each; an extra work suit, $15; six pairs of sweat socks, $6; gloves, $1.50.

At the hardware we bought knee pads, $6, and across the street we picked up steel-toed boots for $19. Then it was into the five-and-ten store for a lunch bucket. I was about to buy a black plastic lunch pail with a Thermos nestled in the lid, but upon learning that I was going into the pits the woman at the counter vetoed the idea. "You'll thirst to death with that bitty thing," she said. "Take a regular miner's bucket so you can carry a decent amount of water. My man carries a Thermos of coffee with him separate, but you daren't go without water."

I bought the round silver miner's bucket with a tray that fitted inside the top for food, leaving the bottom half for water. The miner's belt and helmet would be provided by the company from the company store and deducted from my first check. Our expenses for equipment were over $100, and the clothes would wear out quickly.

The first morning of safety school, we packed the kids into the car and Grada dropped me off at mine #37, where the school was held. Class began at 9 A.M., and we had time for a leisurely breakfast at the cabin. Grada was excited and relieved that I was on the company payroll before the baby's arrival. Now miners' welfare insurance would pay the entire $600 cost of delivery.

Fifteen of us sat in a small room at mine #37. Two of the group were former miners who were returning after driving trucks for a few years. The rest were mostly teenagers and men just out of the service.

We watched films on how to treat electrical shock, resuscitation techniques, and the location of various pressure points that would stop bleeding. The instructor passed around an emergency respirator that we would carry on our belts underground. An old safety inspector dropped by and told stale jokes, and the company paymaster explained the pay scales and deductions. Although most classes went underground, arrangements for our group were botched, and we never left the classroom.

September 19

THE FIRST DAY UNDERGROUND. THE ALARM WENT OFF AT 5 A.M. I dressed quickly in cold, damp clothes, built a fire downstairs, and made coffee. When the fire was crackling, filling the room with an orange glow, I went back upstairs and woke Grada. She came down half asleep in her heavy blue bathrobe and cooked eggs, sausage, and toast.

After we had eaten, I went back upstairs, swaddled the kids in a heavy blanket and carried them down. Grada took a flashlight and my lunch bucket, and we went across the bridge to the car. I returned to the cabin and brought back the mining clothes. In what was to be our daily routine, we drove off in the darkness to mine 18-D.

Most of the road on the fifteen-mile trip was ravaged by potholes and ruts that shook the old Chevy's frame. We passed through River Gap, a quiet village of wide lawns and Victorian houses. Then, in a dark valley outside of River Gap, we saw the mine tipple's insectlike superstructure silhouetted by rows of greenish lights. The coal that comes out of the mine on belts is dumped at the tipple. There it is sorted, cleaned, and loaded into trucks and railroad cars for shipment. We drove to the bottom of the hill and turned toward the portal, the mine entrance.

At the portal, the parking lot, covered with a rufous shale called red-dog, was crowded with pickups and four-wheel drives. I climbed out of the car with my bucket and clothes and kissed Grada good-bye. We had decided that if the baby came while I was at work, Grada would drop the kids off with an old friend, Mrs. Hart, a seventy-six-year-old woman who helped raise me and now lives alone in Greenridge. Then Grada could call me from Miners' Hospital.

As I watched Grada leave, I remembered that when I was a child, my sister and I had played in the abandoned strip mines about a half mile behind my house. We were forbidden to enter the shafts, but we often went inside anyhow. I recalled one winter afternoon when a half-dozen of us were fooling around near a small abandoned shaft. I was about six and my sister was eight. Icicles had formed inside the shaft, and the sun shone directly through them, creating a scintillating series of diamond-like stalactites. My sister insisted on going inside despite entreaties from the rest of the group. After a few minutes she emerged. The following day we passed the shaft again and saw that it had caved; it had simply disappeared under tons of rock.

I went inside the big metal washhouse, hot and crowded with men dressing at long green benches. Large heaters in the corners blew hot air to dry the filthy, crusted clothes that dripped from wire baskets hoisted to the ceiling. Newspapers covered the coal dust on the cement floor. Behind the benches, bright, thin chains ran from a pipe through pulleys in the ceiling to the baskets. The lamp man assigned me a basket and showed me where it was.

After dressing in long underwear, two pairs of heavy socks, rubber boots, canvas trousers, a sweatshirt, work shirt, canvas jacket, gloves, knee pads, and a new, shiny red helmet, I was drenched with sweat. I went out to the hall and put on my lamp, sliding the light into the notch on the helmet and strapping the battery onto my belt. On the other side of the belt hung the self-rescuer, an aluminum can with a mask inside to filter out poisonous fumes for a half hour in case of emergency.

I stood with a group of other new men by the corner water fountain near the bosses's office. The waiting room was an L-shaped hall between the washhouse and the cage—the elevator—at the far end.

A boss came out, took me by the arm, and led me over to a barrel-chested middle-aged miner. "This is George the timberman," he said. "Stick with him."

George nodded and looked away. He was seated on a bench with a group of other older men. There was no room for me on the bench. I returned to my bucket by the water fountain, but I kept an eye on George, ready to follow the instant he stood up.

I was nervous and sweaty and wanted to talk to somebody, but the tenseness in the air was forbidding. Most of the men either hunkered next to the wall or sat on benches. Some walked, tested their lamps, spat into garbage cans, or knocked against the floor with their hammers. Bosses appeared from the office, looked around, and ducked back inside. The hundred-odd men on each shift are divided into crews and each crew has a boss. In charge of all of the crew bosses are two or three section bosses per shift, who report to the mine foreman. He reports to the mine superintendent, the ultimate boss.

Night shift came up, faces black, eyes glazed. Twenty of us at a time replaced them in the cage and dropped a couple of hundred feet. At the bottom, we went down a concrete ramp into the main shaft and climbed into the mantrip, five open rail cars with a big electric motor at each end. A plank was laid across the inside ends of the rail cars for seats.

When the mantrip was full, the bosses climbed in. The motorman clanged his bell twice and we rumbled out. We had to duck our heads in sections to miss the overhead rails. Beams from our lamps bounced along the gray sides of the shaft. After five minutes or so, George flagged the motorman with his light. We climbed out and stepped back while the mantrip rolled past. Then George crawled back to a hole about a yard high. I followed him through a trapdoor. He picked up a wooden wedge about a foot long and handed me one.

"Cap pieces," he said. "Makes it easier walking."

Those were his first words to me. I wanted to ask a lot of questions, but he took off, moving briskly, using the wedge as a third leg through the shaft. I hobbled after him with crablike lurches, smashing my head against overhead rails and timbers and scraping my back against roof bolts. My light kept falling off. Finally we stopped, and George took the light off my helmet, dented the bracket that the light fitted into, and snapped the light firmly back into place. We went on. My leg muscles knotted. Sharp pains shot through my back. Sweat ran down my face and I was panting hard. We stopped beside a pile of logs about seven feet long and a yard around.

"Find a piece of wire. We have to drag these through that trapdoor up there," George said in a slow, mild voice.

He found two pieces of wire, tied one around a prop, and handed the other one to me. Then he left, dragging the prop behind him. I tied my wire around another prop and pulled. It wouldn't move. I turned around, dug my heels in, and heaved. The prop slowly plowed after me. Stopping every few yards, in about ten minutes I made it to the trapdoor. George was waiting. We returned to the pile and dragged the rest of the props, then stacked them beside the trapdoor.

After a short rest, George said, "Better hold your hat goin' through here. Strong wind."

He opened the door and air whistled through the darkness. My legs quivered, and I was still gasping.

"Go on, go ahead!" he yelled, holding up the door.

I closed my eyes and plunged through. A howling stream of wind whipped my helmet off. The light snapped off and cracked me on the nose. I was in darkness, eyes squeezed shut to keep the dust out, until George climbed through the door. When he closed it, a hollow boom echoed down the shaft and the wind stopped. I put on my helmet and we sat on a small pile of props.

"We'll take a rest here. It gets a little tight up ahead," he said. He took a snuff can out of his pocket, removed the lid, and aimed his light on the pocket watch inside. "Almost nine."

My lungs gasped for air. My legs shook. And we had been inside only two hours. I decided the hell with it; I would do what my body could do, and when it quit, I quit.

A few feet beyond the trapdoor was a rusty steel three-wheeled cart. We loaded the props on the cart, then George pulled and I hung onto the back and tried not to let him haul me with the props. I struggled to keep my back down, but roof bolts—steel rods put into the roof for support—scored my back in long gashes. Finally we came to a caved-in section where we could stand.

"This is it," he said. "This roof's bad, you better sit over there."

The roof had caved in in sharp angles up to about fifteen feet, leaving jagged, heavy rocks hanging in half-fractured steps. A constant cold wind blew through the shaft. Water rained from the roof. It was impossible to stay dry. My back ached from banging along behind the cart and my trousers about the boots were soaked from stumbling through scummy black pools along the way.

"It's cold here," I said. "Must be a lot of fun in the winter."

George was measuring the height of the roof, using a long, thin stick. He motioned me over to help him saw a prop.

"The fan draws 130,000 cubic feet a minute through these shafts," he said. "Closer you are to it, stronger the wind gets. Come winter, if it's zero outside you'll have thirty below around here. We're close to the fan."

He raised the prop, stuck a wedge in each side, and tapped them with an ax.

"Gets pretty severe. You got to keep moving," he said. He told me to go back, then drove the wedges tight with the ax. He wasn't dressed half as warmly as I was; he wore a torn plaid shirt and thermal undershirt covered across the top with bristly gray hair. He didn't seem cold, but, as he said, he was always moving while I sat there.

I was thirsty, but our buckets were hung on props back by the trapdoor. George said they had to be hung up to keep the rats away.

He set props the rest of the morning. I sat along the side, out from under the rocks, and watched. Once in a while he would shove a rock bar, a long crowbar, into a gap and bring down thudding shelves of rock.

At eleven thirty we went back to the trapdoor and ate

lunch. Grada had packed three meatloaf sandwiches on homemade wheat bread with mayonnaise and red, ripe tomatoes, a bag of chocolate chip cookies, and a bag of sweet white grapes. The water in the bottom of the bucket was ice cold and tasted of the spring. George said he didn't eat much underground. He had two plain bologna sandwiches on soft white bread.

After lunch we had to drag more timber and props through the trapdoor and back to the work area. I crawled and dragged and pushed the cart. Chunks of skin came off my back each time I scraped a roof bolt. Finally the day was over, and we went back to catch the mantrip.

In the washhouse I hung my caked, dripping clothes on the hooks attached to the bottom of the basket. In the showers, a lot of backs sported red, raw scrapes and scratches, so it wasn't something I'd learn to avoid. The steaming water really stung the cuts when it hit.

Back at the cabin, I sat on a rocker on the porch and watched the tops of the pines wave in the breeze. The kids played in the yard while Baron pranced around with a stick.

Grada brought me a Jack Daniels with lots of ice, then sat with me, laughing at the way I looked.

"You'll just have to build up different muscles for working under a yardstick all day," she said. "Everybody must have to go through it."

I pulled up my T-shirt and showed her the bloody streaks down my back. My fingers were bruised. Every muscle in my body ached. I told her I was thinking about cashing it in.

"But what can we do? We have no place to go. You have to keep on, at least for a while," she said.

It bothered me that except for a few hours pushing the cart and dragging props, I hadn't done a thing. George had set the timbers while I had sat and watched. Yet at the end of the shift, I wasn't sure I could walk.

During my four years at Penn State, I had been a blond, crew-cut iron freak. At one point I weighed 215 pounds, and could standing-press my own weight. Now, at thirty-four, my mirror reflected a skinny, hairy ex-hippie struggling to convince himself that the basic skeletal structure was still sound and that getting back into shape would take no time at all.

"The baby was kicking all day," Grada said. "At least we're ready for him now. Wait until he's born. Then, if you still can't take it, you can quit and we'll go on from there."

September 20

I AWOKE STIFF AND SORE. IT TOOK A LONG TIME TO DRESS and I had to skip breakfast. Same routine, carrying the kids over to the car and driving over rutted roads in the first tinge of morning light.

I was sitting on the bench in the hot, dusty waiting room off the washhouse when Donchus, the short, thick-necked crew boss of Main C, came over. "You found yourself a home," he said. "Go with my crew to Main C."

There are usually four men in a crew. The operator runs the Lee Norse continuous miner, a thirty-three-foot digging machine with two cleated heads that spin and oscillate into the coal face. Two mechanical arms under the heads scoop the coal onto a chain belt that carries it through the center of the machine and out to a tail where it is dumped into a buggy, a twenty-five-foot shuttle car that rams into the back of the Lee Norse and takes the coal that streams out on the clattering metal conveyor. The buggy runner, or shuttle car operator, drives the buggy to the belt and dumps it. The belts take the coal out of the mine to the tipple, where it is cleaned and sorted. The operator's helper pulls the power cable out of the Lee Norse's way, helps set rails and planks as work progresses through the seam, and hangs canvas to direct the air for ventilation.

In rooms, or sections, that have been mined by the previous shift, the roof bolter, or pinner, bolts the roof to the rock strata above it with steel pins every four feet to make the roof safer. The roof is rock, sandstone or soapstone, or sometimes slate that breaks up easily. Under good roof, planks can be set every four feet. If the roof is breaking up and threatening to collapse, king rails that weigh ninety pounds per foot are set every two feet.

I was the helper. Remick, thirty-two, a short, stocky, reserved man, ran the Lee Norse. Larry, a wiry farm kid, ran the buggy. When I sat with the crew in the mantrip, no one spoke for the entire cold, bouncy ride. We crouched under the rails and dodged the streams of water from the roof. When we reached Main C, we walked back a long, black, wet sidetrack off the main shaft. Everyone wolfed down a sandwich before unloading the supplies from rail cars. The supplies were sent up on the belt to the work site.

The belts, yard-wide, trough-shaped rubber strips that bump long over rollers, are the basic transportation system in the mines. Their main function is to carry coal out at five hundred feet or better per minute, but at the beginning of each shift they are reversed and slowed to half speed to carry men and supplies to the work areas, a trip that takes a half hour or more in some places.

"Get in and throw some of those rails out," Remick

told me. I climbed into the rail car and grabbed the end of a rail. I couldn't move it.

"Here, look out," said Larry. I moved aside, and he picked up one end and threw it on the edge of the car. I tried the next one, but it wouldn't budge. Embarrassed, I tried jerking it as he had. I bent my legs and straightened my back as shown on the diagrams on the safety bulletin board. The rail might as well have been welded to the car. Larry motioned me aside and threw the ends of four more rails onto the edge of the car. Then he and Remick slid them out. Each took an end and tossed them on the belt.

After loading the supplies, including thirty four-foot props and thirty bags of rock dust (white powdered limestone, fifty pounds each), we climbed onto the belt. I laid low and kept my head down while we rolled along two feet under the roof. The belt slanted sharply in some places. In others, it was slack and jerked along until it tightened again.

When we reached the tail end of the belt, we unloaded the supplies and walked to the face of the coal seam where the Lee Norse sat. Larry showed me where to pump grease and oil into the Lee Norse while Remick knocked out the safety props set up by the previous shift.

My job was to keep the electric cable and plastic water pipe away from the machines. The mine machinery runs on electricity broken down from 12,000 volts. If either Remick or Larry ran over the power cable, it could explode.

Remick hung his respirator, a face mask with a cotton filter, around his neck and turned on the Lee Norse. Larry went back to his buggy, and suddenly the air was full of black dust and the shattering, grinding noise of the Lee Norse. I sat in the darkness, sweating in the dusty, airless heat. There is supposed to be at least 11,000 feet of air a minute directed onto the face, but the amount changes each time someone hangs a piece of canvas and redirects the flow. After each cutting, black dust obscured Remick's light even though he was only a few feet away. I couldn't see the other side of the fifteen-foot-wide shaft. I listened for the warning whine of the buggy over the roar of the miner, ready to pull the cable back if he reversed, up if he moved ahead. The buggy slammed right into the back of the miner, and Larry didn't watch for me. In many places he had two feet of clearance on each side of his twenty-five-foot machine. If he knocked out a prop holding a rail, the rail would crash down on the buggy and slide back, possibly taking his head off. Remick said that it had happened to a man in a nearby mine who had been running a shuttle car for twenty-three years.

Toward the end of the morning there was a short break while the belts were not working. Remick shut off the Lee Norse and said to me, "Donchus needs a buggy runner pretty soon. Watch the way Larry handles it and in a week or so you can run the buggy during lunchtime."

Donchus came by periodically. He would hunker by the Lee Norse and watch me shovel coal from the side into the path of the machine. When Remick's patience ran out and he took the shovel from me, Donchus would bellow at me, "You'll have to get off yer ass on this crew."

I would be on one knee, gasping, wet, nodding compliantly, and suddenly he would yell, "Goddamit, start shoveling! Throw some rock dust around!"

When he left, Remick laughed and said not to pay any attention to him, but it was obvious that Donchus had sincere doubts about adding me to his crew.

Remick shut off the Lee Norse while Larry ran the

buggy to the belt. He slumped against the controls and wheezed into his respirator, until the buggy whined toward us. Then he hit a row of levers and the Lee Norse jolted into life, heads spinning and shaking. One of the lobster arms that scooped the coal was broken and I had to shovel the coal from the side of the shaft to the center of the machine so that the other arm could scoop it in. I did my best, kneeling in the water from the sprayers on the heads to throw shovelfuls of wet coal, but my clumsiness with the shovel always sent Donchus into a tirade.

The six-inch electric cable was heavy. The Lee Norse shifted violently, reversed, slammed against one side and the other. I was supposed to grab the cable and haul it backward as fast as the miner chased it, meanwhile listening for the whine of the buggy so I didn't get run over, trying not to trip over bags of rock dust and props. All of this in a space that varied between a yard and forty-eight inches high. My back again was striped with long gashes. The safety visor on my helmet made a callus on my forehead.

Suddenly a massive slab of rock crashed down on the Lee Norse. I was sitting half asleep behind the machine when Remick slammed it into reverse. I dragged the cable as he backed out. Then I was supposed to break up the rock with a sledge while Remick and Larry watched the roof and listened for more cracking. How do you swing a sledge with a six-inch clearance? Remick grabbed the sledge and slammed it against the rock a few times. The rock fell apart. You had to hit it along the grain, he said.

"You think you're gonna like the mines, buddy?" Remick asked me later. It was the third time I had been asked that question, always in the same laconic tone. Remick, at thirty-two, has been in the mines for ten years. He pulled the previous operator off the Lee Norse after a rock had spilled his brains over the seat. Remick was in the seat the next day, and he has been there ever since. I asked him why he didn't find some other work.

"With these continuous miners, your lungs go after five years. Nobody wants you after that. I just got my house sided. Where else can I make twelve thousand a year?" he said.

The roof became bad. Remick and Larry would kneel and watch it crack, with dust sifting through the breaks. Then they would go forward and rap the roof with a hammer. Larry said that if there was a solid ping the roof was good, but if it sounded hollow, like a drum, it was bad. Remick backed out frequently and we set rails every couple of feet, sliding them from the buggy up onto the Lee Norse, then onto the heads. The Lee Norse lifted the rail against the roof, then we measured the height underneath, cut props, and wedged them tight. The work never stopped. After a rail or two was set, Remick was back on the Lee Norse, driving into the face again. The coal that I had to shovel because of the broken scoop was soaked from the sprayers on the heads. I kneeled in the water and tried to use my back as Remick showed me, but he could never wait. He was always aware of how many buggy loads Larry had run to the belt. When he spoke of the job, hatred and bitterness filled his voice, but he wanted to turn in the highest buggy count at the end of the shift.

For the last few hours I stumbled around half blind. Sweat poured down my face. My knees were like hamburger from the coal inside the knee pads. My fingers were stiff and swollen from being caught between a couple of props. I couldn't feel the burning gashes down my back. My body had finally turned off the pain, and it took every ounce of will for me to drag the cables around. At the end, I really didn't care if they exploded or not.

"You'll get used to it," Remick said. "It takes a couple weeks."

Finally it was over. Safety props were set in several rows up to the face to keep the roof from collapsing before permanent supports were put in. It was the helper's job, but Larry and Remick pitched in while I bungled around. The roof height was measured with a saw and a wedge. Then a prop was marked and cut. The prop was stood on end, and wedges were driven in with an ax. It looked simple. But the bottom and roof were never level, and it was hard a cut a prop straight by the light from a headlamp. We tried to avoid cutting the props by finding high spots, then stacking wedges underneath the props and banging three or four on top, an illegal practice covered by smothering the bottom of the prop with coal. Bags of rock dust were split open and the white dust was thrown against the sides and roof to improve visibility and keep the coal dust down. Everything was finished in a great rush and everybody ran for the belt. I scuttled after them desperately, clanging my head against planks and rails, scraping more flesh off my back, crawling through the low sections. After riding the belt out, we waited ten minutes for the mantrip. There had been no reason for the mad dash except the excitement of getting out.

We went up in the cage and into the washhouse. Some men were half undressed before they reached their baskets. Then we went into the showers with a bottle of shampoo and a bottle of detergent. Three teen-agers just out of high school whom I had met at safety school were nearby. I asked them how they were doing, and they said not too badly. One of them noted that I didn't look well.

"You're on the best crew in the whole mine, though," he said enviously. "Those guys are coal-hungry mothers."

Outside, it was cool fall afternoon. Men rolled their windows down and headed out onto country roads for a great nose pick; clumps of black, gummy coal and viscous snot flicked out the window leaving fingers, fingernails, and face smudged. Eyes were ringed with black grime, miners' mascara.

By the time I reached the cabin I thought I was dying. My body sung with fatigue, a physical hum that seemed to vibrate through me. Grada made a drink for me and tried to cheer me up. I sat on a rocker on the porch while the kids begged me to romp on the lawn with them.

"Remick told me the boss wants me to become a buggy runner," I said. "There's a little space on the side of the shuttle car where the buggy runner jams himself in. He has to duck his head under the rails, watch where he's going, and go like a bastard with a two-foot clearance on each side. I'd be dead in a day or two."

Grada laughed. Now that we could have the baby free, nothing bothered her. "You'll get used to it," she said.

"No, I won't," I insisted. "Donchus, the crew boss, is a slave driver. They're the best crew in the mine, full of pride and confidence. A crack team. I think Donchus suspects that I've been assigned to his crew as some kind of joke. He isn't amused."

"Give it a little while, at least," said Grada. "Wait until the baby is born. Then we'll see."

September 21

My knees ached unmercifully. I kept thinking that this would be such a simple job outside. Several times I just couldn't force my knees up and had to crawl with the cable. I narrowly missed being run over by the Lee Norse when Remick suddenly backed out. The Lee Norse spewed dust and coal all day. At times I couldn't breathe. Sweat filled my respirator. Kneeling in the water and shoveling wet coal made my elbow snap every time it was extended. Donchus was constantly yelling at me to shovel faster, put the canvas up, horse that cable back, all the while sitting his fat ass on his hammer. During lunch, I took the buggy for a run along the straight part of the shaft and narrowly missed bringing several rails down. I couldn't fit into it.

However, I may yet be saved. Rumor has it that a deal has been made: Donchus' son used to be the buggy runner on his crew until he bid on another job. Now he wants to return to his father's crew, which would allow me to escape.

September 23

Zachary was born today in Miners' Hospital. Free. Grada's contractions started early in the day, so I stayed home. We waited until they were a few minutes apart, then drove to the hospital, a large, brown, depressing building six miles away. Inside, the nurse at the desk told us that doctors were on duty only at certain times of the day. I asked what happened in case of fibrillation or cardiac arrest. The nurse said that she and the other nurses were trained to handle it until the family doctor arrived. It was like this in most rural areas, she said; there was not enough money for doctors.

I was not allowed in the labor room except for short visits. Grada lay on a metal bed surrounded by curtains in a darkened room painted a pale, institutional green, lit dimly by a night light over the bed. A cord with a button on it hung from the wall. A nurse let me stay longer as the contractions became stronger. Grada held my hand and kept her eyes closed. She didn't talk much. Sometimes the contractions woke her. She would squeeze my hand very tightly. Her eyes would open wide, but she would not cry or make any sound except sometimes a little grunt. The nurse came in, dressed like a mechanic in green coveralls, and offered pain medication. Grada refused with a small shake of her head. She wanted to have the child naturally.

By the time she was fully dilated, the doctor had arrived and we went into the delivery room. Grada was frightened and asked several times if everything would be okay. I said sure; after all, babies were born all over the world every day. It didn't seem to comfort her much.

Although neither of us mentioned it, we were wondering whether the baby would be deformed from X-rays. While we were in Spain, Grada had been thrown from a horse when she was a month pregnant, and had broken her shoulder. The Spanish doctors had taken X-rays over and over again, then botched the setting job. Later, they took new X-rays, broke the partially mended bone, and reset it.

Now the doctor and nurse positioned her on the table and strapped her arms down. Then the nurse pulled Grada's knees up and strapped her legs into stirrups. Everyone was dressed in green gowns and face masks. The lights had a deathly cold, bright glow seen in cheap bars in the early mornings. At the bottom of the table, a large lamp spotted a small, brilliant beam on Grada's cervix.

The strongest contractions of all came, and the doctor urged, "Push! . . . Push!" and a little brown patch of hair, the top of the baby's head, appeared. It drew back after the pushing stopped.

"Just a few more and we've got it," the doctor said.

Grada's face changed from a dark, flushed red during the pushes to a pale white afterwards.

"Now! Push! . . . Push harder! Again! That's it, good! Quit pushing now. Pant. That's it, pant."

The baby slid out slimy white and blue, squirming and bawling. The doctor drew some of the mucus out of the baby's mouth with a tube, then wrapped a small blanket around him and put him in a box. I counted his fingers and toes and put my finger in his hand.

"Is it good? Is he all right?" she asked me.

"He's perfect," I said.

She was shaking and trying almost successfully not to cry. The doctor finished stitching and threw the clamps into a bloody dish.

"Congratulations," he said to me. "You can go home now."

On the way back to the cabin I stopped at Mrs. Hart's to check on the kids. I hated to burden her with the kids, but she was the only person I could find who would take them. The kids were fine, playing on the floor with mementoes that Mrs. Hart had saved from my childhood.

I spent the rest of the afternoon trying to borrow heaters. The cabin is no place to bring a newborn, but I have no paycheck. The Greenridge bank refused to give me a loan without a cosigner.

September 25

THE WEATHER HAS TURNED COLD, ESPECIALLY IN THE swampy basin around the cabin. I tried sleeping with no heaters to see if the kids would be able to stand it when Grada came out of the hospital. With all the covers from the kids' bed plus ours, two pairs of socks, a T-shirt, and a sweatshirt, I shivered all night.

Something has been leaving droppings on the bridge and around the cabin, and entered the cabin last night. Baron, upstairs with me, woke up roaring and barreled downstairs. He returned with nothing.

October 5

OUR NEW SON, ZACHARY, LIES CRYING IN HIS CRIB. WE ARE back at the cabin, all of us, with two large electric heaters in the bedroom. Our bed is separated from the kids' room by an old frosted-glass partition that was taken from my grandfather's office in town. It doesn't reach the ceiling, so the heat passes over the top.

This was the last night of the hoot owl shift, 11 P.M. to 7 A.M. I was sent up to the long wall, the most dangerous place in the mine and the main coal producer.

The long wall is a seam of coal about 470 feet long and four feet high mined by a heavy cleated wheel called a shearer, which is the height of the seam. The shearer whirls along the face of the seam and dumps the coal onto a pan line, a metal conveyor belt behind the wheel. The roof is supported by about 150 chocks, hydraulic jacks with ironing board tops connected to the pan line by metal shafts. The row of chocks extends the length of the wheel's run. As the shearer mines the face of the coal, the chocks are rammed forward. The roof behind is left to collapse. If you happen to be crawling through the chocks when the roof gives way, the crash of tons of rock a few feet away behind the jacks can shake you up. A boss from another mine once told me in the post office that men on the long wall crew had a problem when they wanted to bid on other jobs, because they had learned to ignore cave-in warnings.

We had a meeting with a young, enthusiastic safety inspector named Rasko before going down. These sessions are held infrequently on different shifts to remind the men of the hazards and to try to impress them with the need to work carefully. Rasko, the inspector, is unusual because he is an avid, energetic, dedicated man who sees the mines as a great opportunity for a young man to get ahead. He is full of enthusiasm about new machinery and technology. Some inspectors are older men who smile their way along, greeting the miners and poking around shafts that have somehow been cleaned up and freshly supported just before the inspectors' arrival.

Rasko told us that we probably knew that mining was the most dangerous occupation in the country, but did we know that we were working in one of the most dangerous mines in the United States? I didn't know that. During his talk, the men sat stolidly along the wall, spitting snuff into tin cans. Some of them closed their eyes. They seemed to have heard it all before and were thankful for the few minutes they could sit on a bench and be paid for it. Afterwards, going down in the cage, I thought about what Rasko had said. I was going to the most dangerous place in one of the country's most dangerous mines for the night shift, the most dangerous time to work.

I was sent to the tailgate with Bobby, a nice, quiet twenty-year-old whose job was to blast off the end of the wall with dynamite so the shearer could move in to begin another pass along the wall. The chocks use soapy, oily water to maintain their pressure. Their hoses leaked, and at the tailgate the roof leaked. We had about a foot of cold water to work in.

Since the roof was low, we had to kneel in the water to shovel the coal that Bobby blasted off. He knocked off about a ton per shot which we shoveled onto the pan line. A very strong, cold wind blew up there. We bored six or eight holes at a time and packed them with dynamite. Bobby explained that it wasn't too important how many sticks were used as long as it did the job. He sent me down the chock line. I squeezed under leaky hoses and crawled under the ironing boards while he prepared to shoot.

The first two blasts went well. Then I crawled away for the third shot. There was a terrific blast. When I crawled back through the heavy, acrid smoke, Bobby was shoveling madly through a pile of coal. He had miscalculated, and our lunch buckets were buried. When we dug them out, his was blown apart. Mine was intact except for the bottom part that had held water. We had not eaten yet, so Bobby had no food or water for the rest of the night. I offered him half of my sandwiches, but he said he didn't eat much underground anyhow. He showed me his bucket: a small drugstore of Pepto-Bismol, antacids, and other remedies. Like most miners, especially the older ones, Bobby is a mess inside with ulcers, hemorrhoids, and nervous disorders.

The shearer broke down often, and there was nothing for us to do but wait until the mechanics fixed it. Once while we were waiting I asked a new young fellow on the chock line why the long wall was the most dangerous place in the mine. He said he didn't know. Later, as he was crawling through the chock line, one of the advancing chocks caught his foot and removed three toes.

By the end of the shift, my knees ached with exquisite pain. The long, hot shower helped, but this was the last night of hoot owl and the fatigue caught up with me. All week I have snapped at Grada, spanked the kids, beat the dog. For some reason, I could not sleep more than four hours a day.

Back at the cabin, I tried to eat ham and eggs before going up to bed. I nodded sluggishly over the plate, dazed and red-eyed. Grada kept up the small talk, chattering about one of her girlfriends who lives in Florida and wrote to Grada to ask if she knew what to do for cervical erosion. I recommended a little contour plowing. Finally, I stumbled up to bed, hung a towel over the window, crawled under the blankets, and sank immediately into sleep.

October 6

TODAY WAS SUNDAY, OUR DAY OFF. I HAD A HACKING cough and lower back strain, but I was not as bad off as last week. We went apartment hunting in expectation of my first check. We found nothing.

It was warm and beautiful at the cabin. The kids, shepherded by Baron, played in the woods. Grada and I discussed how much longer we can last at the cabin. Our main worry is of Zachary catching cold or pneumonia. The cabin has actually become home. The polished wood floors gleam and the big fireplace of heavy railroad stone heats the living room at night. One cold water tap and an outhouse have become minor inconveniences, excepting for using the outhouse at night. I always approach the outhouse slowly, waiting for the yellow light inside which I turn on from the cabin to disperse the menagerie of raccoons, woodchucks, spiders, three-inch dragonflies, and giant crickets that use the place as a club after dark. The animals leave when the light goes on, but the insects remain, flattened against the walls and ceiling, long antennae gently waving. It bothers me to enter there at night, drop my pants, and sit over the jagged hole in the planks, atop an insect Calcutta.

From the accident list on the washhouse bulletin board: Two men were working on a power box moderating 12,000 volts. One tested it with a screwdriver, and the box exploded. One man was in serious condition in the Pittsburgh Burn Center, the other was in similar condition somewhere else. Notation: Turn off current before working on power box.

A rock fell on a man's hand and broke it. Notation: Test roof and sides before entering area.

One of the young men who started with me had a finger smashed while throwing crib blocks. Notation: Better communication among workers.

Plus the kid on the long wall who had his foot caught. At first, I was relieved to see how many old-timers looked uninjured. That was before the half-hour lunch breaks when they talked about broken backs, smashed bones, and hundreds of narrow escapes. Ganiak, a slim, ulcer-plagued face boss my age, told us about a man who was running a Lee Norse when the power went off. He decided to use the time to change bits on the digging heads. While he was at it, the power came back on. He had forgotten to shut off the Lee Norse. A bit caught his pant leg, pulled him under, and the machine ate him alive.

Another man was riding the belt at high speed, five hundred feet a minute or better. It was illegal, but no one will scuttle like a crab for a half mile under forty-two-inch roof when he can hitch a ride on the belt. A small cave-in ahead allowed the belt to pass through, but not a rider. He slammed into it head first and was killed instantly.

The United Mine Workers national contract expires November 12, one month away. Talk in the washhouse indicates a strike of at least a month.

October 13

FINALLY WE HAVE ENOUGH MONEY TO MOVE OUT OF THE cabin, even though it is into a trailer. Everyone talks about how much colder this year is than the past five have been. The last two weeks at the cabin, we had to put food into the refrigerator at night to keep it from freezing. Zack caught a cold. The fireplace turned perverse and belched smoke into the room instead of sending it up the flue.

The trailer is on a hillside among a hundred or so others. A coal miners' trailer camp with work suits drying on clothes lines and old cars mounted on milk crates; the kind of place you might see from a train window and wonder fleetingly what life is like there.

There is a bedroom at each end of the trailer with a tiny bathroom, the kitchen, and the living room in the center. The furnace is between the bathroom and the kitchen with the result that the bedroom on the far side, Hadley's and Bo's, receives no heat. The walls are brown plastic masquerading as wood. In the living room, a dirty, mustard-yellow carpet covers the floor. There is a sofa and one chair. Our bedroom has two beds, the kids' room has one. We borrowed a cot from the cabin for Bo. Zack sleeps in the bathroom with his crib wedged between the bathtub and the toilet. It is impossible to get to the sink on the other side of the crib without picking up the crib and backing out with it, which wakens Zack.

Monthly charges: rent, $135; oil heat, around $50; cooking gas, $10; garbage pickup, $4; cable TV (no reception without it), $5; electricity, $10; telephone, $6. Total: $220 a month.

From my paycheck, based on $42 a day with Saturdays time-and-a-half, initiation dues of $100 for the union are deducted off four checks, $25 each. Also deducted is miner's clothing from the company store, which is so low-profiled now that it's stuck off the road behind mountains of boney, the slag that is separated from the coal and piled in heaps outside the mines. The store, a small, stone building with a gas pump outside, charges about a third more for its merchandise than other stores, but purchases are deducted automatically from a miner's check and there is no dunning during strikes.

I can remember as a child being taken by my father to the old company store that was in the center of Cokeville, an old-fashioned place with barrels of Louisville Sluggers, knee pads, shirts, and other merchandise stacked all over the wooden counters and hanging from the ceiling beams. In the old days, miners had to buy from the company store. If they didn't buy enough to satisfy the company, they were fired. In hard times, miners often received no money on payday; their checks were simply deducted from the company store's bill. I remember when the old Tennessee Ernie Ford song "Sixteen Tons," with the line "I owe my soul to the company store," was banned from the local radio station by coal company pressure.

Cokeville remains an archetypal coal mining town. Rows of company houses abut the street and piles of boney surround the valley. The boney piles used to burn, giving off a cheery orange glow on clear nights. On rainy days, however, a foggy, sulphurous haze with the stink of rotten eggs hung over the area. The boney piles remain, with the eerie desolate periphery of dead trees and vegetation around them, but the piles don't burn anymore. The company houses are privately owned now and some have additions added to the bungalow frames.

October 21

MY HAND WAS RUN OVER LAST NIGHT BY A ROOF-BOLTING machine. I had been sent as a pinner helper. The man I assisted was not a regular roof bolter, and we had an old, outdated machine that went over my hand before either of us knew it was moving. I was on soft coal, so no bones were broken. A patch of skin was missing, but it didn't bleed, and I did not come out.

During lunch, one of the crew mentioned the time a few years ago that a man was guiding cable into a winch and stepped into a loop. The loop snapped tight around his leg, and the only way he could save himself from being pulled into the winch was to twist his leg off at the hip. Donchus, the crew boss at Main C, pinched off the man's arteries and held them in the ambulance all the way to the hospital.

More talk in the washhouse of the coming strike. The bargaining over the national contract that expires November 12 is not taken seriously by any of the men here. Everyone is getting ahead on his car payments, stocking the freezer, making plans for hunting camp.

The ride in the mantrip has become very cold. The shaft is so low in some places that we have to crouch almost to the level of the car. Icy water runs from the roof, always finding an ear or an open collar during the trip.

At home, things are difficult. Zack is in Miners' Hospital with a 103° fever. Grada refuses to wean him and makes the six-mile trip four times a day to breast feed him. She said the hospital is so poor that there are no blankets for the baby and he has to be covered with a towel.

October 25

I WAS ASSIGNED TO THE SUPPLY MAN TONIGHT. HE USED a long, low battery-driven machine called a kersey to haul supplies to the face where the crews were mining coal.

Loading the kersey was tough. For example, one load was a hundred bags of rock dust (fifty pounds each), six rails (half ton or more each), and six planks (several hundred pounds of splinters). The trip took about a half hour. The kersey jolted and crashed through the ruts and potholes in the shaft. The kersey runner, a big, genial man named Joe Morgan, told me to keep my mouth open. Morgan, my age, is 220 pounds of muscle, a steady worker who has missed only one day in the past two years.

There were no shocks in the kersey, Joe said, and the drop into a deep hole could crack a tooth.

The kersey had a bright headlight that cast sparkling reflections off the translucent fungi growing in all colors over the planks and props. Pools of water reflected in our light down the shaft and through the white, furry moss overhead. The sides and roof were damascene with wet rock dust patterned in whorls from the Lee Norse's spinning heads. It was like riding into a crystal paperweight fairyland. We surprised a couple of chubby rats in a crosscut and chased them, but they got away.

I waited alone while Joe went into another section to discuss supplies with a boss. After he had gone, the shaft was silent as a tomb. I turned my light off and was immersed in darkness so complete and quiet that I felt like a child scaring himself in a closet. When I flicked the light back on, I noticed shapes far down the shaft which seemed to dissolve when I turned the beam on them. Sometimes they floated along the edge of the light. I experimented, keeping the beam fixed on one point, turning it on and off. I was wondering about what battles were fought down here a few million years ago and what graves we were desecrating.

When Morgan returned, I told him I was seeing ghosts. He laughed and said sure, it gets foggy down here, too.

October 28

DAY SHIFT. I RISE AT 5 A.M., SORE AND BLEARY-EYED. Lately the sky is steel gray, and cold rain and mist make driving difficult. I dread seeing the greenish lights on the tipple when I crest the hill. My elbow is bursitic and snaps every time it's extended.

My knees ache. It's like working under the kitchen table for eight hours in icy water, shoveling coal and hoping the roof stays up. There are easy days, setting timber in a high, dry, warm place with good roof, but for every good day you know a couple of bad ones are coming.

I was sent down along the belts to shovel out the rollers and the mounds of coal dust under the belt. I worked on the wire side where bundles of cables hung next to the side. The belt spun past and the roof was low, about a yard, which left no room for the shovel between the belt and the roof. If the belt caught the shovel, the shovel would fly back into my face. A boss whom I hadn't seen before instructed us to shovel with the belt running away from us. He said if the belt caught the shovel, let the shovel go. Then, when we were at the work site, he told us to start working in the opposite direction against the belt, in knee-deep muck. In a space cramped by cables, low roof, and a fast belt, the biggest danger was having an arm or hand caught in a roller.

I thought that it was illegal to shovel next to a running belt, but to make an issue of it was to go on the mine foreman Cooper's shit list and spend the rest of my days under bad roof and in water. As foreman, Cooper reports only to Kurtz, the superintendent.

Cooper is short and fat with a jaunty, almost mincing, walk. His habit of stroking his pencil mustache and scowling at the floor when asked a question could be mistaken for befuddlement and belie his reputation as a martinet with a nasty temper. His absent-minded look can change in a flash. When angry, his jowls quiver, his pale, round face turns crimson, and his voice slides up the scale into a loud screech. He is despised and feared, and reportedly never forgives a real or fancied affront. The men on Cooper's enemies list cannot always quit and go to another mine, either. The blacklist is still in use.

So, we worked. The coal was wet and heavy. Because of the low roof, the shovel had to be extended to arm's length and twisted so that the coal would fall onto the belt. My lamp cord continually caught in the hanging cables. The rescuer and battery on my belt hooked onto the belt support cables. The dust that flew off the belt clogged my nose and eyes. I was angry, and I became angrier as the shift progressed. I told myself that if I were not deep in debt, I would walk out and tell them to shove it.

During the lunch break, I asked the two new young men working the other side of the belt what they thought about our situation. They shrugged. Both of them were just out of high school and were not disposed to make trouble in a job that was paying them $12,000 a year. They couldn't have cared less that they were risking injury or death on an illegal job.

The shift dragged. In jobs that were simple drudgery a man could daydream while he worked, but here a moment's lapse of concentration could be dangerous.

I was still angry when I left the washhouse. Outside, it was clear, sunny, and warm. On the way home I stopped at the River Gap Hotel for a beer. Zurko, slightly drunk, was sitting at the end of the bar with a crowd of other men. The air, blue with smoke and filled with pizza odors from the ovens in the back, was stifling. Shafts of sunlight slanted through the windows and flashed off the dried rings of soda and beer on the tables.

Though Zurko and I had talked about getting together,

I hadn't seen him since the Greenridge festival. I suggested that we throw a couple of six-packs into his truck, take a few of his beagles, and go into the woods to let them chase some rabbits. Instead, we headed for Greenridge and continued drinking. I had mentioned my feelings about shoveling the belt all day. Zurko said, "You're the same as the rest of them. Sheep. There's always an excuse to let the operators kill you."

In Greenridge, he launched into a denunciation of the Catholic Church while we drank. All of us had gone to the parochial elementary school. It was, and is, the only elementary school in town, and the Catholic Church is the only church. It was Zurko's idea that the church conditioned us as children to believe that life was supposed to be pain and suffering, and that if we endured it without complaint, the Lord would reward us on the other side. He said that our schools were the coal operators' training ground.

To me, the influence of the church was minimal, although it was not until after high school that I decided that people who wore funny clothes, had nothing to do with the opposite sex, and talked to statues were not the ones to be guiding my life.

Zurko contended that the Catholic Church was to blame for repressing a child's freedom to question and for teaching him that a life of degradation was the admittance ticket to heaven. I said the school system was more at fault; the kids weren't taught anything practical to make a living with, so that after graduation they had to either pump gas or go into the mines. The bartender, a stout, tough woman named Georgia, said the church *was* the school and to quiet down or get out.

We continued drinking through the night. At closing time, 2 A.M., the shades were drawn and the overhead light put out. Six of us, all miners, remained. Zurko and I talked about our childhood and debated whether it had conditioned us for the rest of our lives.

Zurko brought up an incident of a few years ago. A young man had come home to Greenridge to visit his parents. He had brought his wife and babies along, and they all camped outside of town in a trailer. A few of the local boys were boozing it up and decided to have some fun. They went out to his trailer and started running circles around it with their car. Finally, they came too close and knocked the trailer off the supports at one end. The young man came out with a rifle and fired at the car. That upset the men in the car, so they ran over him, breaking his leg.

Around 7 A.M., I left the bar to go home. The battery in the Impala was dead. Enraged, I attacked the battery with a hammer and demolished it, then returned to the bar and left the car in the middle of the street. When I came out again, an hour later, the car had disappeared. I went to a friend's house, a mechanic-welder, and found the car in his garage. He had seen it in the street and towed it. He suggested that I go to the garage and buy a new battery.

I walked to the garage and spotted a new demonstrator on the lower floor. I decided on a quick nap in the back seat, and awoke completely disoriented some time later during a test drive by a prospective customer and the garage owner, severely startling all of us.

November 2

IT WAS COLD AND DARK AT 5 A.M. THE CAR BARELY STARTED, even with the new battery. My breath fogged the windshield. In the washhouse, I discovered that someone had stolen my helmet liner. I had to steal someone else's. After suiting up, still bleary-eyed, I went into the hall and read the accident list. More of the same: smashed fingers, sprained backs. I looked around the hall at the men hunched silently on the benches and I wondered what difference it really made that we could now afford motorcycles, snowmobiles, boats, pickups with C.B. radios, and all the other toys. We were still working ten hours a day, including travel and suiting up, six days a week, living with coffee in one hand, a candy bar in the other, glancing at the clock every half hour.

November 8

All four Nova Coal Company mines went on strike today, four days before the scheduled national UMWA strike. We went out early because of an incident at a local union meeting in Cokeville a few days ago. I wasn't at the meeting, but the men in the Greenridge Legion said that a group from mine #32 had come to the meeting drunk and upset over $1.95 in overtime owed to one of them. They took over the meeting and threatened to kick anyone's ass who refused to support them. To avoid a fight, John Yunko, the local president, adjourned the meeting without a vote, which meant that the other three Nova mines had to go out in sympathy. That cost us today, which is Friday, tomorrow, which is a time-and-a-half Saturday, and Monday, a paid holiday. We are not paid for holidays when we are on strike. One might think that this whole scene had been engineered by the company, but it's just an example of miners screwing themselves.

November 9

THIS AFTERNOON GRADA IS AT THE LAUNDROMAT. THE kids are outside playing, except for Zack, who is asleep in the crib in the bathroom.

The neighborhood: on our left is a young couple named Cutler, matched like bookends, both small and thin with glasses. When they are outside with their baby they hover over him constantly. The kid is going to think the sky is plaid—every time he looks up while crawling he sees the front of his father's shirt.

Above them on the hill is a fat woman separated from her husband. She has three daughters, a thirteen-year-old and two younger ones around Bo's age. We seldom see her, but when her husband visits we can hear screaming arguments.

Next up the hill are the Patricks. The father, an ex-fighter, is a foreman at Bethlehem Mines. Somehow, eight people live in their trailer. On nice days the father and the oldest boys, one fifteen-year-old and twins about twelve, drag weights out of the trailer and set up a gym on the lawn. Then the father trains the boys, watches them spar and skip rope and do bench presses. They seem to be a loving, close-knit family.

Across from them, directly above us, are Larry Bills and his wife, Jane. Larry, short, stocky and balding, is a twenty-nine-year-old mechanic at Nova #37. Jane is a loud, energetic, thin brunette with two daughters, six and four, by a former marriage. Sometimes in the late afternoon I sit with Larry in lawn chairs he sets up in front of their trailer. We drink beer and survey the camp life spread out around us.

On our right, the first two trailers in descending order on the hill are occupied by young couples with babies.

The young man in the topmost works in a supermarket. The one below him is a miner at Bethlehem.

Below us, around the bottom of the hill, are retired folks who take the camp seriously. Their trailers are surrounded by tiny picket fences, and the immaculate yards are planted with shrubs and trees. Two of the trailers have carports attached. These old folks intend to die here, while the rest of the camp consists mostly of young couples working toward a mortgage.

The camp teems with kids, and like kids anywhere they make a lot of noise. Unlike kids anywhere, they are well disciplined. We all work hoot owl every third week, and every kid knows what Daddy is like during that week. Not just *his* daddy, but *every* third-shift daddy. Pillows smother the telephones, blankets and sometimes tarpaper black out the bedroom windows, and any kid foolish enough to ignore a warning to be quiet or to go play somewhere else makes the mistake only once. Miners' children organize their lives around their fathers' hoot owls. If a kid wants permission to do something, he arranges to ask for it during day shift or second shift. During third shift, kids stay out of sight as much as possible.

November 12

THE NATIONAL STRIKE FOR A NEW CONTRACT BEGAN TODAY. No one expects it to be settled for at least a month. In fact, most of the avid deer hunters among us would be mad as hell if a contract was settled before then.

Grada and I, however, are already in money trouble and it won't be long until we are back in hock to Master Charge.

November 14

GRADA HAS FOUND A BRIDGE CLUB IN ARCOLA. THE CLUB is mostly middle-aged, educated people, many of whom knew my parents. When we meet in the street, there is always polite chatter, sometimes with proposals for getting together. But these invitations never materialize. Anyhow, I am more comfortable with miners. Grada is more at ease with the bridge club.

November 18

A GORGEOUS, WARM, SUNNY AFTERNOON. I WAS SITTING in the Greenridge Legion playing the old spinet piano and swapping stories with Ernie, a friend who grew up down the block from me in Greenridge and now lives thirty miles away. Ernie, short and swarthy, has always had unlimited energy along with an innate sense of life's absurdity. We were in the navy in Hawaii together, he at the Barbers Point Naval Air Station and I on a ship in Pearl Harbor. For a month or so, while my ship was in dry dock, I had had a job as a pit drummer in a strip joint in the Chinatown section of Honolulu. It was a native club with an all-black band, except for me. Between shows we played hard rocking blues and the girls worked the tables out front. Ernie used to come down to the club, and those days always formed a part of our fund of old stories when we got together.

The Legion's back room had a yellow hardwood dance floor and murals on each side wall. The rear wall was glass brick that sparkled in the late afternoon sun and filled the room with hot, bright light. I was playing a few coal miners' folk songs I had made up when Zurko and old Chernovak came back from the bar to join us. Chernovak, a big, stooped miner with gray hair and a grizzled beard, always calls me Doc, after my father. Chernovak has been plagued with a bad back all of his life and made frequent visits to my father. Once when I was a boy my father told a story at supper about him. Chernovak had been in that afternoon about his back problems, and my father decided to take an X-ray. He sent Chernovak down the hall to the X-ray room, one of a series of treatment rooms. "There's a stack of green gowns in there on a chair," my father told him. "Take one and put it on so it opens out the back, then press the button by the door and I'll come back."

After about twenty minutes there was still no signal, so my father went to see what was up. Chernovak had all fifty gowns spread out over the floor, the X-ray machine, the chairs, everywhere. He gave my father a helpless look and said, "I dunno what the hell's the matter, Doc, every damn one of these opens out the front."

While I was playing, Zurko and Ernie got into a discussion about the public's ignorance of the violence in modern coal miners' and operators' relations. Most people, said Zurko, thought that violence had gone with the thirties and forties and that now negotiations took place at a bargaining table. Ernie was a meat salesman, and like most people, even from a mining area, he knew little about life underground. Zurko had some back issues of the *United Mine Workers Journal*, the union magazine sent to miners every two weeks. He was talking about the Brookside mine strike of last year in Harlan County, Kentucky. In June of '73 the miners at Brookside had voted to be represented by the UMWA rather than the company-oriented Southern Labor Union. In late July, after Duke Power Company, owners of the Brookside mine, had refused to sign a labor contract, the miners walked off their jobs and went on strike.

By August '74, the strike was still on. Duke Power hired groups of goons and called them security guards. The guards rode around at night firing automatic weapons into striking miners' homes. On August 9 they attacked the home of the union local president, Mickey Messer, and fired about a hundred rounds into his house after midnight while Messer, his wife, and four children, including a fifteen-month-old infant, hid behind the furniture.

Zurko had another report, a comparison study showing that European mines were much safer than U.S. mines, even though European mines were deeper and more dangerous to work in.

Ernie passed the report to me. It was testimony by J. Davitt McAteer, the UMWA Safety Director, and titled *Safety in the Mines: A Look at the World's Most Hazardous Occupation*. The study said that six times as many Americans died per million man shifts worked as did West Germans, four times as many Americans were killed as English miners, and three times as many Americans as Poles. Each year one out of every fourteen American miners was injured.

The study went on to say that European mines were safer despite the fact that they were normally several thousand feet deep, while American seams were only several hundred feet below the surface. The deeper the seams, the bigger the safety problems—more methane gas was liberated, fresh air problems increased, and roof falls were more frequent and came down with greater force.

I didn't know if Ernie cared about mining problems, but he and Zurko had played football together in high school, and they had always been good friends. Finally, Zurko asked me to play some Mantovani arrangements. I told him that wasn't my style.

December 8

I WENT TO THE LOCAL UNION MEETING THIS MORNING IN the crowded, one-story green cinder block union hall in Cokeville. A stairway to the left led to the bar. Ahead, through the cigarette smoke, the wide stairs to the dance floor were jammed. Under the sign NO FIGHTING—AUTOMATIC SUSPENSION, men jostled and moved inside to tables, searching for friends. I sat at a table with Zurko and several other men. Zurko asked if I was celebrating the Feast of the Dubious Assumption today. One of his weird Catholic jokes.

At the far end, behind a long table, sat John Yunko, the local president, a tall, thin man with white hair and a plaid tie. He was flanked by two local officers. John rapped his gavel and yelled in his booming voice, "How come I never see you guys except when you wonder if you're going back to work?"

Somebody suggested that more men might show up if meetings were held in the afternoons instead of Sunday mornings. That brought a laugh. Even in the mornings there was a constant flow to and from the bar. As the meetings dragged on and frustrations grew, the tendency increased toward settling a point with a punch in the mouth.

John was in trouble. A lot of men were angry over the incident that had caused us to go out three days early, the drunks from mine #32 who had taken over the meeting.

After scolding us for not attending meetings and complaining about the loss of solidarity among the brothers, John discussed the national contract which had been ratified last week. Our local had voted against it, 377 yes to 391 no. John and the others who had been against the proposed contract were especially angry because the TV in the bar had been announcing imminent ratification when most of our local hadn't voted yet. Men would be at the bar having a few beers, making up their minds, and see the newscaster reporting celebrations in Washington, D.C. over the ratification.

"Hell, they could put a whore in every crosscut and it wouldn't satisfy those guys," an old timer next to me muttered.

John went over the contract briefly. He didn't think we had gained enough, although it amounted to a fifty percent raise in wages and benefits over the next three years. The largest gains were in benefits. The old contract paid $150-a-month pension. Under the new contract, pensions went to at least $200 immediately and at least $300 a month to miners retiring in 1976 at the minimum retirement age of fifty-five with thirty years in. What interested most young miners were wages, the least of the increases. An immediate nine percent raise the first year and three percent each of the next two years would boost the average daily wage from $45 under the old contract to $54 at the beginning of the third year of the new contract. I figured that with overtime, working six days a week, most miners around here would earn between $14,000 and $16,000 in 1975, depending on how many walkouts we had. A cost-of-living allowance, something the auto and steel workers had all along, would add as much as $24 a week over the next three years.

Other benefits included $100-a-week disability for up to a year, five days' sick leave versus none in the old contract, three more paid holidays making twelve in all, and a new $75-a-year clothing allowance.

I hung around after the meeting and had a few beers with some of the older men. Mitch, the man who pushed my application, raised a dozen kids in a tiny house during the fifties and sixties. Harry, a squat, fast-talking old timer, returned to the mines two years ago after trying to make it as a salesman. Both men have almost thirty years underground.

Mitch was scheduled to go in at 11 P.M. hoot owl, the first shift back. There would be water everywhere. The grease and oil in the machines would be frozen. Cleaning up, pumping out the water, greasing the machines and getting them to run would be a terrible way to start night shift in this weather. The more Mitch talked about it, the more he drank until he was too drunk to go in.

"They talk about solidarity," he said. "Hell, we won't ever see that again. These kids don't know hard times. They don't know what to do with their money."

We talked about football for a while, and then just sat there and drank. Finally Harry said, "I'll tell you one thing, I never have to wear a watch underground now. I can tell you when it's two hours to quitting time right on the nose. My legs give out. They knot up behind the knees. One of these days there'll be a fire drill and we'll have to walk out. I won't make it. It gets me worse every week."

Word came in from the bar that the construction workers and truckers hadn't settled yet. Miners have never crossed anybody's picket lines. Christmas was around the corner and we had been out for a month with no income. No one knew when things would be settled and we would go back to work, but suddenly men were buying shots with their beers and smiling.

December 12

I WENT DRINKING LAST NIGHT IN COKEVILLE AT BARLEY'S bar. Barley is a tough, sinewy brawler who has maintained a stable of girlfriends, married and single, since I was a child. He's over fifty now, lined and weathered from a life of construction and ditch digging, still full of vigor and still with several women on call. I asked him how he did it. He said, "You got to keep active. Stick and move."

Later, when I was maudlin and drunk, we talked about my father. Barley had known my father as a young man, and he told me stories about him.

"He was the same as you," Barley said. "He bounced around four or five colleges in his time. Always raisin' hell. But he got to be a doctor, though, and here you are in the mines. It's harder to be a fuck-up and get by these days. Why, if you had money back then, like your dad's family, hell, you could get away with almost anything."

I regretted that my father and I hadn't been close. He hadn't been one for fishing and ball games, but we literally never did anything together.

Later, Zurko and I were playing pool on the small table and talking about hippies. To Zurko, the term hippie is synonymous with worthless asshole. I told him some stories about my four years as a hippie in Spain. Once, when I was returning from one of several trips down Morocco to Marakech, I thought it would be cool to smuggle some hash into Spain. The sentence if caught was six years and a day. The six years weren't too bad, the hippies said, it was that last day that got you.

In long, blond hair and beard, big, dark glasses and a floppy hat, driving a rusted Volkswagen squareback, I was invisible in the load of stoned-out freaks that stumbled off the ferry in Algeciras, Spain. The unfortunate on foot had to go through personal inspections, including the spread-your-cheeks routine.

Those of us in cars formed a half-dozen long lines. Sitting on fenders or hoods in the hot sun, we watched the Spanish *guardia* in green ex-Nazi uniforms and mirror sunglasses dismember the cars in front piece by piece, knocking on all compartments with rubber dolly hammers and tossing door and rocker panels around. Meanwhile, another *guardia* on a platform scrutinized the rest of us for signs of panic. This psychological warfare went on for about an hour until I reached the head of the line.

I was smugly content, having had the cleverness to stash my dope in the glove compartment. There was only one bad moment while they were breaking down my car. I had left the car registration in the glove compartment underneath the patty of aluminum-wrapped hash. When the *guardia* asked for the registration, I was hard put to be nonchalant about blocking his view while I opened the glove box. It worked, though, and I stopped intermittently along the next five miles of Spanish coast to celebrate with other drivers whose cars were parked in various states of collapse after the *guardia*'s half-hearted efforts at reassembling them.

Zurko loved these stories. I think it reassured him to find people crazier than he was.

PART 4

December 23

Our district finally went back to work today, the last miners in the country to return. The construction workers had settled and the truckers called a two-month moratorium.

I was driving to work in a snowstorm when a brand new pickup stopped abruptly in front of me in the middle of River Gap. With poor rubber on the Impala and ice on the road, when I hit the brakes it was like putting the car into neutral. The choices, to be made in two seconds, were to swing into the left lane and force the car coming the opposite way to decide what to do, to pile into the rear of the truck, or to swing right, up onto the sidewalk, and try to miss the trees. I chose the sidewalk and a tree went right through the bumper and radiator. The crash gave me a hell of a jolt and spread my lunch bucket all over the floor. I put the lunch bucket together and went out to join the curious crowd examining the smoking car. Then I climbed into the pickup with three beefy young guys who worked at the same mine. We were jammed on top of each other in a sweaty, stinking bundle. Nobody said much about the accident. We were all thinking about the misery of the first day back.

At the portal, I called the garage in Greenridge and asked them to tow the car. The washhouse was hot and dusty as ever. Heaters in the corners were still blowing hot air through the baskets of filthy clothes overhead. Superintendent Kurtz, a tall, slim, blond man with a military bearing and a calm, direct manner, held a brief meeting before we went down. Kurtz, a former cop, had a reputation as a cocksman when I was young. At 18-D he is as adept at handling the men as his foreman, Cooper, is inept. Kurtz, as superintendent, is in an autocratic position, but as top dog he has more to do than prowl through the shafts checking on his men. That is left to the short, rotund Cooper, second in command, who spreads nothing but hatred and contempt wherever he appears.

We stood in a circle while Kurtz spoke from the center in his thermal underwear like a football coach. Take it easy, check every part of every machine before energizing, check the roof and sides, don't take chances.

The elevator shaft is also an air shaft. When the door opens at the bottom, freezing air blows hard down the cement ramp to the main shaft. During the six weeks we were out, winter settled in. The fifteen to thirty degrees lower temperature underground, depending on nearness to the fan, put real teeth into the wind. The water pouring from the roof which we had ridden through in the mantrip was now a gauntlet of icicles. Anyone in the mantrip who looked back instead of forward was punished by a sharp crack against his helmet.

I worked with a fat, greasy kid named Riley on a supply job. Riley did as little as possible, putting the burden on his buddy. Underground, buddy means coworker and not necessarily friend. Since he was senior man on this job, he was responsible for what was done. I decided to do what he did and no more.

Before we got started, he said, "You know, everybody tries to help everybody else out down here, but you got to look out for yourself. If something happens, like the roof starts working, I'll tell you right now, I'm long gone."

There was no need to worry. By the time we had one trailer loaded with planks, props, and rock dust, he decided it was lunchtime. After lunch, he said he never felt like doing much after eating. We didn't take a single load to the face. The first day back, said Riley, you could always claim brake problems with the kersey.

Every so often his leg would cramp and he would howl and hop along the shaft. Most of us had put on weight and lost the muscle tone needed for low work. We ate lunch with a belt worker who said he had been battling rats in one of the mechanics' shanties. "Poor little fellers starved for six weeks," he said. "No wonder they were eager."

The cold wind penetrated deep into my head. I noticed that many of the old-timers had stuffed cotton in their ears. I had a headache when I got home. I told Grada about the car and she cried.

Later, I sat in the living room chair with a drink and we talked for a while. We knew what we said was not important; it was the communication that would hold us together.

"The news last night said the economy is worse," she said. "It doesn't look very good for us."

"Well, we still have a little time before Hadley starts to school. Maybe we can save some money by then."

"We haven't saved a dime yet. You hardly get through

a month without a walkout or two. What are we going to do when you get hurt?"

"I won't get hurt," I snapped. "I work carefully. Not everybody gets hurt."

"That's not what I read. If you stay there long enough, you'll get hurt. Or your mind will go, if it hasn't already. What about your friend when the ice fell on his elevator? That's enough to shatter your nerves."

She was referring to an accident at another mine. I didn't know the men on the elevator, but I had been drinking in River Gap with Zurko when one of the men who had been in the cage was describing the experience. The mine was one of the deepest in the area, about a thousand feet. Fifteen men were halfway down in the cage when a chunk of ice the size of a refrigerator broke off at the top of the shaft. They could hear it coming—boom! *boom!* BOOM!—and it smashed onto the cage. Four steel bars across the top saved them from immediate death, but the blow broke three of the six cables suspending the elevator. Lights and communication went out. The ice block hung through a jagged hole in the roof. The men in the cage swung very gently, knocking against the sides of the shaft, five hundred feet up, listening to the wind howling past for an hour before they were rescued.

"You're not the person I married anymore," Grada continued. "When you started in the mines you were afraid and nervous, but at least we shared it and talked. Now you've turned off so well that you can't turn on again. You're either a zombie or a maniac."

"You've changed, too," I said. "That big martini every afternoon doesn't help your conversation. You yell at the kids a lot. Anyhow, you can hear men going nuts almost every night somewhere in this camp. It's an occupational hazard."

"Would you like to know what I'm afraid of?" she asked.

I nodded.

"I'm afraid we'll be here forever. I can see it happening and I can't stop it. It scares the hell out of me, and I can't stop it."

At that moment Hadley dumped my lunch bucket on the floor in the kitchen. She is always fascinated by the water in the bottom half and the metal insert on top that holds the food. I had forgotten to pour out the water, and it was all over the floor.

December 25

CHRISTMAS. I BORROWED AN ANCIENT PONTIAC FROM PAT Lanoski, the mechanic-welder who towed my car when I was drunk. It doesn't start very well, but it will have to do until I can find another heap.

This morning Grada and I awoke at 6:45 to the sounds of rustling papers and childish whispers. We stayed in bed listening to Hadley's and Bo's hushed, ecstatic exclamations in the living room.

Finally we dressed and went out to watch. Bo, wide-eyed in his blue flannel nightsuit, sat festooned with bright ribbons in a pile of wrapping paper while Hadley showed him how to raise the bed on his new dump truck.

We lit the tree lights and ate breakfast. Then Grada opened her present, a new bridge book. My present will wait until we hit better days. We shared the kids' exultation and I spent the rest of the morning trying to assemble a plastic train set whose directions were lost in the pile of papers. Hadley paraded in and out of her bedroom giving us a fashion show of clothes made by Grada's mother.

December 26

THIS MORNING AT 6 A.M. THE OLD PONTIAC DIED AS I WAS driving to work. I pulled off the road in the darkness and cold rain, wondering what to do. It was freezing. I had no flashlight. There was no telephone for miles, and little traffic on these country roads at this hour. Then I remembered that I had put the new battery from the Chevy in the trunk of this Pontiac. I went back and opened the trunk and found the battery with a broken set of jumper cables.

Standing in the freezing rain, I peeled back the insulation on the cables with a penknife, then hooked the cables around the battery terminals and jumped inside the car. It worked! I threw everything in the trunk and went on to 18-D feeling invincible.

After work, at home in the trailer with rain still drumming on the roof, we had a talk about money. Grada had a budget that showed we could still pay off our debts and possibly save a little by March or April. Walkouts and bad luck considered, it was not a realistic plan, but we had to have a goal.

December 31

NEW YEAR'S EVE. A TIME TO AVOID FUN SEEKERS, SO Grada and I stayed home and got sozzled alone, watching Guy Lombardo (or a replica of Guy Lombardo) wear his funny hat. And at the moment of truth, as the ball on top of the Allied Chemical Tower descended slowly to snuff out another year, rather than brooding on the future I found myself recalling an incident in the past.

In my youth, the teen-age club met in the Cokeville union hall. One night when I was fifteen, a live band was playing there and during intermission I went up to fool around with the drums. When the band came back, the drummer, who was the band leader, offered me a job, and I accepted.

My first club date was at the Greenridge Legion for New Year's Eve. It was hot and stuffy and jammed with people. Each time some of the men passed by the bandstand, they put a mixed drink on the bass drum for me. By midnight I was draped over the drums, completely plastered. The band leader took me home and told my father I would never play with his bunch again. The next day, in the midst of my terrible hangover, my father sat on the edge of my bed and said to me, "I want you to understand something. Most people in this town are fine, good folks. But there are a few who like nothing better than to see you or me disgrace ourselves. You will meet people like that all of your life, so keep what I tell you in mind, and be careful."

He never again referred to that warning. Although I disgraced myself often enough in the passing years, I continued to play the drums in local clubs and later through four years at Penn State. Then I decided that drums were too much trouble to set up and carry around, and quit them in favor of the piano.

My mother, a graduate of the Cincinnati Conservatory of Music, once played for Flo Ziegfeld in his New York apartment, and later played for Earl Carroll's *Vanities*.

When I was six and seven she begged me to let her teach me, but I refused. My sister took lessons and I copied her pieces by ear and played them as though I were reading the music, thinking myself clever. Over the years, I became proficient enough to beat out a rambling, boogie honky-tonk that would attract girls at parties, but I would never be good enough to turn professional.

January 1

RAIN FOR THE PAST TWO DAYS MELTED ALL THE SNOW. Then a miniblizzard dumped three inches this afternoon. Wind rocks the trailer. The front door is sealed all around with heavy duct tape, as are all the windows.

Last week Pat Lanoski wanted his car returned. I had no time to look for a new one, nor, without the Pontiac, a car to look with. Then I heard that a young fellow at a nearby loan company, Jack Bartlett, the kid brother of a girl I went to grammar school with, was selling his car for $500. It was more than we could afford, but we needed a car immediately, and since Jack was employed by the loan company there would be no trouble arranging a loan. I told him I couldn't take time to check out the car. All I wanted was his word that it was dependable transportation. He assured me that it was, and put through the loan.

If anyone else bought a car in this fashion, I would be among the crowd of chuckling head-shakers passing remarks about naïveté. The car drives at an angle down the road. The frame is out of line and the motor isn't worth a damn. When I called him at the bank, Bartlett said, "Well, you aren't getting your money back. That's it."

What to do? No sense in serving time for assault and battery. I just keep going to work; there is no time for anything else.

January 2

Another hoot owl this week. I was sent to a mined-out section last night to remove fiberglass pipe with Richard, twenty-seven, a chunky, blond five-year veteran of 18-D from Tippleside.

Mined-out sections are dangerous because they are abandoned. Three or four inches of icy water covered the bottom. It was too low to walk, and if we crawled we would have been soaked for the rest of the night in a strong, cold wind. The roof was three feet high in most places, and cracked everywhere, with splintered planks overhead and rails twisting out. On the way up we had to slosh through a long, stinking bog in the middle of the shaft with sticky muck on the bottom. We both slipped into it and caught water in our boots.

To separate the pipe, we had two chains and a jack. We plunged our arms down into the icy water to put the chains around the pipe on each side of the joint. Then we wedged the jack between the chains to split the joint. After jacking the sections apart, we had to drag them, duckwalking, two hundred feet to a crosscut, where we knocked a hole through a brattice, a cinder block wall. We pushed the pieces of pipe through the hole so that later we could stack them on the other side and someone with a kersey could drag them out.

Then we went back to take another section. With no place to kneel or sit, my knees ached terribly. Water spilled from the roof everywhere and the biting wind stiffened our joints. At 3 A.M., to get out of the air stream, we crawled back to the crosscut where we had come in. We stacked some of the pipe and sat down, hoping to catch some sleep. Richard said that a few weeks earlier he had been setting timber, working alone in the same section farther down. He was trying to knock out a prop, he said, but it had been too tight. Finally he decided to saw it out. He had sawed about halfway through it when he had to crawl away to answer the phone. He had just picked up the receiver when a couple of hundred pounds of rock suddenly caved on the prop he had been sawing.

"I missed bein' killed by ten seconds!" he said in his high, strident voice. "I left the fuckin' place. Somebody else can take that fuckin' shit out."

He said he planned to be married next June, but he wanted to be out of the mines first. His nasal voice twanged with resentment on anything but his favorite subject, the girls he was screwing. Richard compared his exploits with what he saw on the silver porno screen in Tippleside. He claimed he should have been making his own movies.

We stayed on the pile of pipe, teeth chattering, shivering, for an hour. A long scrape on my back which I had received from a hook in the roof burned. Finally we couldn't stand it any longer and we went back to work to keep warm.

January 3

A BUGGY RUNNER WAS HURT LAST NIGHT. HE HAD MADE too tight a turn with the shuttle car and it pinned him against the rib, the side of the shaft. Buggies are twenty-five feet long, and the turns into crosscuts are right angles, often with roof so low that the buggy runner has to hang sideways and try to judge the turn by flashing his light from one side to the other. Under rails, it is vital to watch the other side because if the buggy runner knocks out a prop the rail will smash down on the buggy and sometimes the driver. Evidently this man had just broken or bruised some ribs.

Richard and I were sent with Pumper to replace a 150-pound yellow pump. Pumper, a garrulous, hawk-nosed older man with over thirty years underground, is always cheerful, ready to help anyone, offering opinions on everything in a loud voice. He takes care of the pumps and flooding. He works seven days a week, but after taxes he clears only $70 more biweekly than I do, and I usually take a day a week off, working five days. Under the new contract, I make $47.50 a day, the same rate as Pumper, and I clear between $400 and $500 every two weeks.

We put the new pump on planks and then onto the belt. The top of the pump just grazed the roof. We went up with it, stopping the belt to carry the pump around the scraper. At the work site, we unbolted the old motor and placed the new one on the plates. I pulled a crib block out of the muck and turned it dry side up for a seat, then took off my gloves for a better grip on the wrench. Rubber gloves are fine for working in water, but clumsy for handling tools. When we removed the gloves, our hands stiffened from the cold, and the tools, nuts, and bolts continually dropped into the black water. Our headlamps focused on the job while our breath plumed around us and we talked about the high price of four-wheel drives and pickups.

When we broke for lunch and sat with our buckets,

peeling plastic baggies from our sandwiches, drinking ice-cold water, Richard regaled us with his latest orgy.

"This girl had a prick in each hand and my buddy was fuckin' her. There's this other one from Slag Hollow, she loves to suck cocks. She don't fuck, she's only fifteen, but she loves to suck you off. There's lots of them in Slag Hollow."

Later, he talked about the Veterans of Foreign Wars club in Tippleside, a dance hall where I played drums in various bands when I was fifteen. The Vets, like most area dance halls, was sometimes the scene of grand-scale brawls. Richard recounted enthusiastically the night a fat man blackjacked an opponent, and another incident involving broken beer bottles and bloody gashes. My general impression was that the bars and clubs were calmer now than a decade ago. It seemed from his stories that fights were fewer but more violent, involving weapons instead of fists.

We left early because Pumper wanted to make the first cage at quitting time. Men in the first few cages are assured of showers. As we walked down the main shaft to the cage, we passed small groups sitting in crosscuts and hiding behind doors with their lights off, peeking out like truant school kids to look for a boss. Superintendent Kurtz instructed the bosses that no one was to quit until quitting time, so everyone hung back until the first light was seen heading for the cage. Then the shaft was suddenly full of men.

We headed straight for the ramp, and men began emerging from all sorts of side shafts and manholes, cavities cut into the side of the shaft for men to duck into when rail cars approached. Suddenly a runaway jeep, an electric car used for personnel transportation, passed us, whizzing down the tracks with no driver. He had bailed out when he found that there were no brakes, and he was running after it shouting, "Look out! Get off the tracks!"

The jeep went another hundred feet and smashed into another jeep that was parked. A tremendous crash derailed the runaway jeep. "Well, that stopped her," said a man in front of me. Some laughter, a few more comments, and it was forgotten.

January 8

SECOND SHIFT. I LEAVE HOME AT 2 P.M., RETURN AT MIDnight. I am Clive the timberman's buddy this week. Clive is fifty-four, with one more year to retirement. He doesn't look like he'll make it. His eyes are slits in a puffed, sagging, toothless face. His shoulders slope to a pear-shaped paunch. In a voice that wheezes with black lung, he mumbles about the arthritis in his knees that prevents him from walking at all in the low. He crawls, and in the circle of my lamp behind him, his haunches shift like an old elephant's. He stops twice between crosscuts, one hundred feet apart, to catch his breath.

While he rests with his back against a prop, legs spread, he gasps and spits snuff at the belt whipping past alongside. His lamp is always on low beam, spreading a dim, diffuse light that forces him to squint even more. I tell him to switch the beam, and he does, but the next time I look it's back on low beam. When he looks down, his lamp falls off the helmet, but he won't let me bend the holder for him. He has lost weight since his wife died nine months ago, and his pants fall down when he raises his arms. He won't buy a new belt.

Clive's life revolves around his tools: the ax, saw, and pick that a timber man needs. In 18-D, notorious for thieves, Clive has stashes everywhere—under stacks of rotted props, behind corrugated metal sheeting, buried under coal along the side. He can never remember exactly where the tools are, and he usually spends the first hour of each shift rummaging through junk all over the mine.

We were assigned to set props along the A-15 shaft, a long, three-thousand-foot trip up the belt. The A-15 belt needs adjustment. It is so slack that it touches the ground between rollers, spaced five feet apart. We were apprehensive about riding it, but there was no other way up except to walk. The long wall boys ride the slack belt every day and look on it as a Coney Island thrill. The five-foot ditches that it crosses, with rocks and water below, are an added touch of excitement.

We climbed on the belt and started it by pulling twice on the jabco, a wire that runs along the roof beside the belt. The jabco is attached to switch boxes every hundred feet. One pull trips the switch to stop the belt, two more pulls start it.

The belt snapped taut and pitched us forward. My helmet scraped the roof. Clive plowed into the next roller with a loud "Whoof!" This continued for about fifty feet until the belt tightened enough to run smoothly.

The two-foot clearance between the belt and the roof was obstructed at intervals by hanging wires, crossbelts, and a scraper, a steel plate hanging diagonally across the belt. Hearing the clatter—*tackatatackatatackata*—of a broken roller ahead was a warning to raise up onto elbows and knees. Going over them flat resulted in a crack in the nuts.

We stopped about halfway up to go around the scraper. The long wall crew usually doesn't bother to stop unless a boss is along. They just hang along the edge of the belt and squeeze past it, which is fine unless the blade catches a lamp cord.

We continued up the shaft at slow speed, about 250 feet a minute. After a while, we saw the red lights that marked the tailgate, where the belt ended. The shaft dipped just before the tail end, and before we saw the red lights, their glow reflected down the shiny rubber trough. Clive, lying about six feet in front of me, heaved himself up, ready to grab the jabco that ran through hooks in the roof. Clive liked to stop as close to the tail as possible so he wouldn't have as far to crawl.

At the tail, heavy steel flanges cover each side of the belt for about ten feet and shield the big roller at the end. Clive pulled the jabco. Nothing happened. He yanked again, but the wire was too slack to snap the cut-off switch. He was still a few feet from the tail. On slow speed, most men could have jumped off the belt. Clive couldn't jump. He started yelling, "Shut the fuckin' thing off! Shut the fucker off!"

At the last instant, just before hitting the flanges, he threw himself off the belt. His helmet fell off. The belt pulled his helmet into the gear end. Clive, screaming, was being drawn into the gear by his lamp cord when I pulled the jabco and heard the click from the shut-off box. After it was shut off, momentum carried the belt a little further. When it stopped, Clive was bellied against the gear. We untangled him, and he sat down on an oil can, shaking and cursing.

"Sumbitch, man get killed that way, cocksuckers don't shut off," he mumbled. He tried to put his lamp back on his helmet, but his hands shook too much. I attached the lamp for him and went back to where I had thrown my bucket when I jumped off the belt. The water in the bottom half of the bucket had spilled. Sandwiches, fruit, and cookies lay in the muck in plastic bags.

After Clive recovered, we crawled back to the work area. It had caved, and a lot of rock was still hanging.

"Motherfuckers, they don't tell you nothin', didn't bring a rock bar or a fuckin' thing," Clive mumbled. A rock bar is a long crowbar used to pull down bad roof. Since we didn't have one, we would have to set cribs and props under bad roof that should have come down. Legally, we could have refused to work there, claiming imminent danger. But refusal meant crawling the half mile out or waiting a half shift until the section boss came up. Then it would be into the office with the superintendent, an inspection by a committeeman, arguments, bad feelings. Clive, in constant pain from arthritis, is dependent on the foreman Cooper's good will to keep him out of water and allow him to spend a large part of his last year working in the high main shaft. Cooper is not noted for his good will. Besides, we had fairly high roof, about forty-eight inches, and it was warm and dry. We knew when the section boss, Hudak, made his runs and when he would appear. Hudak is short and wiry and stutters when excited. He is a nice enough man as section bosses go, and, since Hudak is easy to work for, Clive figured the job could be strung out for most of the week.

On the other hand, the roof was soapstone—shiny, black rock that drops without warning. We would be working on the theory that it wouldn't pick the particular moment that we were under it to collapse, a hope that has accounted for any number of dead miners. At least sandstone usually cracks first, giving a second or two to jump.

After sawing a prop, Clive had to sit for a while to catch his breath. He was harder to understand than usual because the day before he had all four of his lower front teeth pulled. We had been out for a few days, the truckers had been picketing, and Clive had been hoping we'd stay out the rest of the week. He said he wasn't having a lower plate put in because it didn't leave room for his snuff, or the snuff got underneath it, or something.

I asked him if he had ever been hurt. He said several years ago he had been helping to dig out a Lee Norse buried in a cave-in. The roof had collapsed again on the men who were digging out the machine. Clive had suffered a broken arm and shoulder and a dislocated back. "Shoulda seen my fuckin' hat," he said. "Still got it somewheres. Looks like a fuckin' train run over it, all bent to shit."

He was out of work for eight months. His wife decided that he should have an insurance policy. He laughed. "Then she went and died before I did." He still lived in the trailer in a small town outside of Pillartown, and still paid the premium on the $25,000 life insurance policy, though he had no children and had no one to leave the insurance to.

It became automatic after a short time to stay upwind of Clive. His long, gurgling farts drifted through the shaft like mustard gas. Even in the main shaft, where the wind was strong, they seemed to hang on. His only acknowledgment was an occasional, "Like to've shit myself with that one," while he peered at the roof.

We were hard at work when Hudak came up. He agreed that we needed a rock bar, then tapped the roof a few times with his hammer outside of the cave, chalked the board with the time and date of his initials, and went on his way to the long wall. As soon as his light disappeared, Clive and I sat down.

We finished timbering most of the cave and had a

crib built by the end of the shift. Cribs were six-inch-square blocks a yard long built up to form a square yard of support under bad roof. We had quit an hour early. The A-15 belt remained on fast speed because there was no one at the other end at the control box to slow it down. Experienced miners just dove on the belt and flew away, but after Clive's close call coming up, we weren't eager to push our luck. I had heard even some of the long wall crew admit to sweaty palms when riding this belt out.

The alternative was walking, or, for Clive, crawling the half mile out, which included two long water holes. We knelt beside the belt and watched it spin past the cone of our lights into darkness, a long, black chute which took nerve to dive into. Props were set at four-foot intervals beside the belt, and when the belt ran on high speed, the props blurred past like guard rails on the turnpike. If we had to jump, traveling at that speed would guarantee at least a bad injury. Red lights are supposed to warn of approaching hazards, such as the scraper and the crossbelt, or planks too low to pass under, but the electricians do not always remove the lights on time. A wrong guess about which lights are safe to ride through could be costly.

Clive squinted at the belt for a long time, muttering to himself. Finally, he crawled away and I followed. He crawled the entire distance, resting twice in each of the hundred-foot spaces between the thirty-four crosscuts. We reached the trapdoor twenty minutes early, so we turned out our lights and waited. When we saw the row of lights of the long wall crew coming down the belt, we went through the trapdoor and out to the mantrip.

There was always more chatter and laughter in the

mantrip going out than coming in. Peewee, a shuttle car operator, was turned around talking to someone when a thick icicle caught him on the back of the head and knocked him into the middle of the car. He sprawled into the puddles of snuff spit, frozen snot, and half-eaten fruit pies. There were roars of laughter.

As we approached the slope to switch tracks into the other shaft, we hunched down inside the car. Coats were pulled over heads, hands clenched under armpits. The wind whistled down the slope at well below zero and moaned through the timbers along the side. After we rounded the curve on the other track, the wind let up and the chatter resumed.

After the showers, one of the long wall crew asked Bobby, the young guy who had dynamited his lunch bucket, if he was stopping at the River Gap Hotel for a beer and pizza. Bob said he couldn't, he wasn't old enough. With a wife, two kids, and a mortgage, at twenty he was too young to have a beer after work in Pennsylvania.

January 9

I STOPPED FOR A BEER AT THE OLD INN ON THE WAY HOME last night. Visibility was terrible. Snow blew across the road and coal trucks boomed past, forcing cars off the edge. Zurko was in the bar, drunk, expounding on the banality of TV and loss of conversation. No one was listening to him. He wanted to go to Arcola to the Notel Cafe, some bizarre hangout of his. I told him some other time.

More car trouble. I checked the points and plugs and reset the timing. Nothing helped. I have to carry ether for the carburetor.

Clive wasn't in tonight. He was complaining about his knees last night. I was sent to the same place with Knopick, a gaunt, loudmouthed young man. He is the butt of a lot of teasing about his half-starved look and the length of his cock, which must be nine or ten inches. It's funny to see these guys in the showers: big, muscular men with tiny cocks and skinny men hung like ponies. Knopick's wife is about three inches taller and fifty pounds heavier than he. She is known to keep him in line.

It was an easy shift, but coming back we had to face riding that belt. Neither of us liked the idea, but we were damned if we would walk. We finally pulled the jabco and piled on. We went down the shaft so fast that it seemed to me like free-falling out of a plane. Fun, providing nothing went wrong. I stopped at the first set of red lights, forgetting that the planks across the belt had been taken out. We had to walk the rest of the way, about a quarter mile, because we couldn't take the chance of shutting off the belt again. It stopped production on the long wall, and they immediately started looking for trouble.

Back in the trailer, everybody was asleep. The furnace chugged and coughed. Furnaces often explode in these trailers, and the volunteer firemen in the washhouse say that a trailer burns completely in five minutes. With a bedroom at each end and one door sealed, no one would get out alive. The thought haunts me on the night shift.

I went into the kids' room. With no heat, they might as well sleep outside. I tucked the blankets under Hadley's curly head and arranged the overcoat that was serving as Bo's second blanket. Outside, yells and curses and sounds of roaring motors drifted across the snow. Drunk miners were trying to pretend tomorrow wasn't coming. I washed in the kitchen sink to avoid waking the baby by carrying his crib out of the bathroom. The crib barely fit between the tub and the toilet and blocked off the sink.

Grada looked good. Her long, dark hair was fanned over the pillow. But she hates being pawed awake. During the day, the kids are around. On day shift, the one week of every three that the scheduling is right, I come home too tired most of the time. Hoot owl turns me into a maniac.

January 10

BACK WITH CLIVE. WE RODE THE BELT UP TO THE TAIL of A-15 again. No problems. The slack has been taken up. When we got off at the tail, we wanted to send some props back another three hundred feet on a second belt. Clive examined the yellow, yard-square power box with heavy cables at each end that operated the second belt. The controls were on one end: start and stop buttons, a handle for forward and reverse, and a knob to adjust the belt speed.

He pressed the start button and the belt clanked into motion. It ran fine, except that it was forward and we needed reverse. Clive shut it off. He pulled the lever around to reverse, kicked it a few times, and hunkered beside it muttering, "Sumbitch won't go all the way over." He kicked it a few more times, but it still wouldn't quite set into the reverse slot.

"Fuckit, we'll try 'er anyhow," he said, and pressed the start button. The box popped like a phosphorus grenade. Flames and sparks shot through the doors. Black, heavy smoke poured through every crack. We jumped back, then Clive scuttled to the box through the sparks and pressed the stop button.

Smoke spread through the shaft. "Which way's it goin'?" Clive yelled. "Goddam, could've got blinded, goddam thing exploded. Sumbitchin' bastard's fucked up now, Jesus Christ, we're fucked."

If the smoke drifted up to the long wall, they would turn in an alarm and the entire mine would go into emergency procedures. Though 18-D is not a gassy mine, there is nothing more dangerous than a fire underground. George, the timberman with whom I had spent my first day, once told me about a friend of his who was near a burning cable. The plastic insulation gave off poisonous fumes, George said, and his friend spent the next full year in a hospital. All of his hair fell out and he would cough for the rest of his life.

We crawled down the shaft and waited while the smoke dissipated quickly through a crosscut. When we came back to the box, Clive gingerly slid the bolt back and opened the doors. The inside was scorched black. The metal fittings were seared off and the cables were incinerated. Clive shrugged and closed it.

"Fuckin' Hudak's gonna shit when he comes up to-night," he said. "Now we got to drag all them fuckin' props back."

Dragging the props was torture for Clive's arthritis and we took frequent breaks. When we weren't moving, the cold air was very uncomfortable. We had left our jackets hanging at the other end of the belt. After we dragged a few props, we rested on them. My teeth chattered. Clive has no teeth, but when he took out his snuff can to look at the pocket watch inside his hands trembled. Dust swirled through the gray plumes of our breath in the lamp light.

Later, while we were eating lunch, we saw Hudak's light flicker far down the shaft. We went back to the power box and Clive muttered to himself, rehearsing his story. Hudak's light moved steadily toward us and soon we could hear the clank of his flame safety lamp and see the small flame swinging back and forth from his belt. Since Hudak is short, he moved through the shaft almost as fast as he would if he had been walking in the high. He was breathing hard as he approached. He dropped to one knee, panting, and Clive started a rambling, mumbled story about the control box. When Hudak opened the box, he stuttered into a rage. "Fu-fu-fu-fuckin' dumb bastard, Clive, y-y-you can't do any fu-fu-fuckin' thing right," he yelled.

Clive mumbled right back at him. They shouted at each other for five minutes, and I never understood a word. Then Hudak slammed the box shut and left. Clive and I went back to our lunch buckets.

"He didn't say nothin' about goin' into the office, did he?" said Clive, grinning broadly.

We didn't set any props. We just dragged them back to the site. Our night's work would have taken a half hour if the second belt had been running. One man would have thrown the props on, the other would have taken them off, moving along the belt and laying the props end to end.

Clive was anxious about the time. He kept pulling out

his snuff can with the pocket watch in it. Originally, he had wanted to leave an hour and a half early and crawl out, but dragging the props had been too hard on his knees. He still wanted to leave early, but he was determined to ride the belt. We quit at 9 P.M. The mantrip didn't leave until 10:25.

We stopped the belt and piled on. Clive stayed about ten feet behind me. We flew down the shaft, through the first set of red lights. At the second set of lights, I pulled the jabco and we coasted to a stop about thirty feet later. We crawled down to the trapdoor, pulled a couple of cinder blocks from a pile, and sat down. We turned our lamps off so that we would be sure to see anyone coming. In the dark, the decayed orange peels smelled more pungent and the oppressive stench of coal dust seemed heavier. The darkness was almost palpable. Water trickled into pools. Rats rustled through plastic bags behind piles of rotted props.

We sat for about five minutes in silence. Finally, I reached out very gently and put my hand on Clive's leg. He jerked away. We sat in the dark for another few minutes, then he turned his light on and went to a stream running under the belt to piss.

I was waiting for his reaction. What did he think it was? A rat? A homosexual advance? A friendly gesture? He started talking about the furnace in his trailer breaking down, how it had cost him over $100, and about a woman who had him to dinner every few weeks. I had never heard him sustain a conversation for so long. The whole situation suddenly seemed fraught with meaning, but I couldn't figure out what the meaning was, or if I was reading too much into the whole business. When he was done, we went through the trapdoor and walked out to the mantrip.

January 13

ANOTHER GRAY, COLD DAY. SNOW EVERYWHERE. I WAS back with Clive underground. We were sent along the main shaft to timber a gray box about ten feet long and six feet wide. Gray boxes broke down the 12,000 volts of power for the mine machinery. We were going to support the roof around the box and set planks over it. It was bad roof, but we were in a crosscut out of the wind and it was not hard work.

Then there was the metallic skirl of a jeep coming and the clack of iron wheels on the rails. The jeep braked hard to a stop, and Vince, the section boss, yelled, "Come on, boys, get in!"

Vince has a deep, booming voice. Nothing excites him more than a leaking pump or hose. His eyes light up behind the thick safety glasses and his big, gnarled hands wave orders at people. He explores for leaks the way young miners comb the bars for women. Vince never tells the men where they are going or what they are going to do there. This sealed-orders, you'll-see-when-you-get-there approach is used by many bosses. The reason for it, if any, escapes me.

Clive was wheezing, "Goddam sumbitch, gotta hide these tools, motherfuckers won't be here when we get back . . ."

"Come on, goddammit!" Vince roared. We climbed into his jeep and he asked if I was afraid to get wet. The temperature above ground was in the low twenties. I told him I wasn't too fond of pneumonia. We clattered down the tracks to a flood of water that reflected the jeep's headlight along the shaft. Off to the side, a big pump was in a pool of water that we could see rising. Behind the water was a trapdoor with its sill already awash.

"There's a submersible in there," Vince said. "Go in and see if it's running."

Submersibles are small, portable pumps that can be moved easily from one flood to another and dropped right into the water. I took a few steps along the edge of the pool.

"Don't try to save yourself, you're gonna get wet!" he bellowed.

I took the plunge. Icy water flowed into my boots and my feet started to cramp. I lowered myself down to the trapdoor and pulled it up. My knee pads caught on the sill. It took a strong effort to squeeze through. Inside, under planks covered with white and orange fungus, the small pump was humming, almost covered by water. A few feet from the pump, the stinking, oily liquid gurgled and bubbled. I poked my head back through the door.

"The pump's running but the hose is broke," I called to Vince.

He cursed and told me to fix it. I was annoyed that he was standing out there high and dry, yelling at me. I told him that I had no tools and knew nothing about submersibles, and that he had to help me. He leaped into the water, spewing more curses. He squeezed through the trapdoor and pulled a screwdriver from his belt. I pulled the slimy hose from under the water and jammed it over the coupler while he tightened the clamps with his screwdriver. We climbed back through the hole. Clive was still grumbling about his tools. Vince told him to shut up. He switched the pump on at the box by the tracks and handed us a piece of thin, green plastic hose from his jeep.

"That old piece is ready to blow any minute," he said. "Take it off and put this one on. I'll be back later."

He drove off and we went back to the crosscut to hide

Clive's tools and give the pump time to clear the water out. When we returned, the hole was dry. The big outside pump was clear and so was the trapdoor. It took Clive a few minutes to squeeze his paunch through the door. Inside, he flashed his light around and said, "They ain't no other way outta this fucker."

The pond around the submersible had shrunk to a few inches in depth. Our job was to shut the pump off, pull the worn section of hose now throbbing with water off the pump, insert the plastic coupler inside the new green section, and clamp it. Then we would do the same thing with the other end. It seemed simple, but once the hose was cut, the pump would have to stay shut off until we finished the job. If the water flooded over the trapdoor, we would be caught.

Clive, wheezing, examined the pump. "Goddam 'em, this sumbitch—lookey here, goddam hose is bustin' outta here. They give you a job, no tools, we could fuckin' drown in here. They ain't no way out."

I crawled outside and shut off the pump. Back inside, as the water climbed around our boots, we struggled with the coupler that attached the old hose to the pump. I held the hose out of the water while Clive hit it with an ax, trying to pry off the coupler.

Clive sounded like an old steam engine going upgrade. After every few swings, his helmet fell off. He would catch it, meanwhile dropping the ax into the water. Then he would put on his helmet and feel around under the water for the ax. Water lapped at our boot tops when I went out to the tracks and turned the pump back on. When I came back in, Clive was sitting on a pile of crib blocks with his chin in his hands, elbows on his knees. His tired face wore a doomed expression.

We watched the worn patch bulging through the

threads that threatened to burst with each throb of the pump. Finally we decided that we needed a saw. Once we had sawed through the hose, we assured each other, it would be a simple matter to pull it off the coupler and slip on the new hose. Of course, once the hose was sawed, there would be no stopping the water until the green piece was attached and the pump turned on. I went for the saw.

When I returned, the water was again pumped out. The submersible slurped obscenely at the remaining puddle.

"You ready?" I asked Clive.

He flashed his light around, through the wispy threads of fungus hanging from the roof, and nodded grimly. I went out and shut off the pump and hurried back in. He held the hose and I sawed it off. Water gushed in. Clive tried to put the coupler on the green sleeve. It wouldn't fit. He yanked on it with pliers, puffing, cursing. He dropped to his knees in the water and peered into it.

"Fuckin' coupler's still in there," he grunted. "You cut it too far out."

I cut it again, closer to the pump, and the green hose slipped on with no trouble. We tightened the hose clamps at each end and I moved quickly to the trapdoor to go out and turn the pump on. Water was a few inches over the sill. When I went out to the tracks and turned on the pump, the light from Vince's jeep away up the shaft glinted off the rails. I waited for him.

He hauled back on the brake and slid to a stop. "How'd it go?"

"Okay. Clive is still inside. I just turned the pump on."

"I have runs to make," he said. "Finish this up and take that other pump out of A-11."

He left and I went back through the trapdoor. When

I pushed the door in, water poured out. It had risen half way up the door. Clive's back was against the roof and the water was just below his hips.

"The fucker's kinked! It's twisted! Turn the pump off!" he shouted.

There was no time. I sloshed through to the pump and picked the whole thing out of the water, then held it while Clive loosened the clamps and put all his strength into twisting the green hose. Then he tightened the clamps. I put the pump back into the water and we waited with our backs against the roof while the water receded.

We were soaked. As the water level gradually lowered, the cold penetrated our wet clothes. I decided to go home. Clive didn't want either to risk the inevitable confrontation in the office over going out early or to be docked the hour that was left. He went back to the crosscut to hide beside the warm gray box, and I waited for Vince to return from his check of the men in each part of his section. When he came back, I rode out with him in the jeep.

January 14

I CUT WORK TODAY. I GOT UP AT 5 A.M., COLD, DARK, AND looked out of the frosted window at the swirling snow. I made coffee and sat at the kitchen table until the kids awoke at 7:30. Bo, in his blue flannel nightsuit, padded out from his bedroom, rubbed his eyes with his knuckles, held his arms up for a hug, and padded on into the bedroom to snuggle next to his mommy. Nothing interrupts his routine.

Zack, snuffling, scratched at the plastic mattress and cried. I gave him a bottle out of the refrigerator and carried his crib out of the bathroom into the living room.

Hadley awoke. Questions: What are you doing home? Where is your bucket? It is her delight to meet me after work and rummage through my bucket for leftover cookies and gum.

Then Grada was up, dour, frazzled, in bathrobe and slippers.

"What's the matter?" she said.

"Nothing. I just didn't feel like going in today. It's the great malaise, a spiritual hangover."

"Bullshit," she snapped, and slammed the bedroom door.

Dawn, gray and white and cold. Coal trucks throbbed on the highway below. From the trailers in rows up the hill, black smoke puffed from the stacks as thermostats were turned up. The television went on. I sat in a chair with Bo in my lap, thumb in his mouth, content.

Controlled silence over breakfast broken only by: What are you going to do today? Have you seen the bills on the dresser?

It was clear that staying out of the trailer was the only way to have any peace. I dressed warmly, took the dog, and headed for the woods.

During the hike through the woods, I thought about what to tell Grada. First off, no explanation was needed. I had made no promises when we were married. Since then, everyone has had enough to eat and decent clothes. It wasn't my fault that she had picked a manqué for her spouse. It wasn't my fault that life was an affront the instant that one's eyes popped open in the morning. A drive into town past flashing quick-food signs and faded Mail Pouch ads on dilapidated barns, through the tundra of frozen, gray fields smeared black with coal dust, on to the mine to sweat or freeze in water under bad roof; that was the price of getting out of bed. Nothing I could do about it. Same thing with drinking in a bar; always some angry drunk wanting to punch someone because he couldn't reach Rockefeller or a coal operator, or whoever he thought was killing him.

But self-pity was a revolting indulgence. One had to work with life and believe that things would turn out right in the end. A conclusion that went against all the evidence. Otherwise, we'd have anarchy, or worse. Socialism, maybe.

These useless thoughts meandered through my head while Baron, ears up, nose quivering, charged through the woods, ecstatic over a world we slogged through with superior indifference. When he returned, panting, eager for me to join in, I ran with him. At last the afternoon turned dark and we returned to the trailer. All was peaceful.

January 31

TWO WOMEN WILL START ON OUR SHIFT MONDAY. BOTH divorcees, one is nineteen with one kid, the other is thirty-six with seven kids. A plywood partition has been put up between the washhouse and the waiting hall so that the women can't see in. They won't have showers, Superintendent Kurtz says. They will wash at home. To come up after a shift, cold, stiff, covered with muck and black dust, and have to drive twenty miles or better is more than I would do. Besides clogging the drains and fouling the showers in their homes.

There are women in the other mines now. Zurko's mine has half a dozen. He said they made the men nervous and hostile because the men had to do the women's work and look out for them. Beneath all the washhouse bluster about making the women work is a lifetime ethic against standing by to watch a woman unload a rail car of fifty-pound limestone bags or seven-foot props.

Rogoski, a crusty, silver-haired old guy who works the crossbelt below the long wall, is supposed to get one of the women. Rogoski is an odd combination; from his basket, near mine, he often gives lectures about the plight of the "little guy" who is being killed by the rich, and one of his suggestions for reform is to take the rich out and shoot them. Meanwhile, his truck bumper is covered with flag decals and stickers telling whoever is following him to love America or leave it. While we were suiting up, the other men offered him advice.

"Hey, Rogoski, yer gonna have to change her rag, too, ya know."

"Don't worry, you handle your end, I'll do mine," he snarled.

"You got to lick her pussy out, too, Rogoski. Got to keep it clean, buddy."

I was hoping for an easy night. I was depressed, waiting on the bench beside my basket. Cigarette smoke and dust drifted through the washhouse. My knee pads already cramped my legs. The straps around the back of my legs cut the circulation when I was seated.

Clive was out again. I worried about him sometimes, living alone in a trailer with a faulty furnace. Nothing I could do for him, though.

Finally I went into the hall. Knopick, the sack of bones attached to a ten-inch dick, as Vince calls him, was sent with me up to the tail of the A-15 belt to reset rails that were twisting out. Knopick said he didn't feel very good either. Coming down from a caved section a few days ago, he had fallen off a ladder and landed on his coccyx.

Knopick said he intended to stay in the mines until he was fifty-five, retirement age, black lung and all. He had it all planned: the time it would take him to bid on an easy job and how much pension and black lung he would draw.

I was still depressed when the shift was over. My father used to define depression as a bad chemical balance in the brain. When I came out of the washhouse to the red-dog parking lot, I decided to have a few beers and change the chemicals around a little. The hairs in my nose froze in the night air. Frost-covered windshields of cars and pickups glistened under the arc lights.

I turned down the hill toward Cokeville, figuring I would stop at Zurko's house. The roads were clear black ribbons under a bright half-moon that shone off the white mountains of boney.

The Cokeville approach passed several mines, their tipples aglow with lines of greenish lights. In town, the sidewalk in front of the bank stood out. There weren't many other sidewalks in town. Most of the houses along the road fronted the black roadside shoulder. The road split the rows of company houses, stores, and bars and passed the park at the end of town. When I was young, the park had been dedicated on a sunny summer afternoon. The black asphalt swimming pool was filled, and the water was fulvous with sulphur. We swam in it anyhow. It wasn't strong enough to burn the skin, but it stank.

Zurko lived in a white clapboard house on a hill behind the Cokeville Athletic Club, which was the union hall. When I went into his house, his wife and kids were clustered around the TV. I remembered his wife as a pert, shapely blonde. She was ensconced in an overstuffed chair now, wore glasses, and sported a head full of curlers, a mouth full of gum, and a nightgown that showed the tops of white, quivering breasts.

Zurko was in his basement workshop, which he called a den, furnished with pine walls, an easy chair, a tattered couch, a stereo, a workbench with a vise on one end, tools hanging along the walls, and shelves of paperbacks crammed along the top of the bench. He was dressed in a torn flannel shirt, jeans, and white sweat socks, and reading a book by Carlos Castaneda. I took a beer from the small refrigerator under the workbench.

"This guy turns into a crow," he said, waving the book. "I'd like to see him underground. See how his philosophy holds up."

"Maybe his philosophy is that if you write books, you don't have to go underground," I said.

We talked for a while about living a perfect life, the idea that most people knew how to live but couldn't bring it off. Finally Zurko suggested that we go to the Notel Cafe. We went out without a word to his wife. I said good-bye, but she didn't look up from the TV. We took his truck and headed for Arcola. Zurko put a tape into his eight-track and a speaker behind my seat throbbed with violin music, the kind heard in supermarkets. After a while, I shouted to him to turn it down.

"I love that shit," he yelled back. "I pretend I'm in an elevator someplace. No problems."

We went through Arcola, down a steep hill and across the railroad tracks, and over to Route 49 to the Notel Cafe.

Inside, the dining room had a great vaulted, tepeelike ceiling, all gleaming wood, with a balcony and skylights.

Off to the side, behind a stained glass partition, in the center of a dark, kidney-shaped bar, a blind organist played requests while a microphone with a long cord was passed around the bar. Anyone could sing. At our entrance, a girl in a white uniform and blond bouffant was aping Patsy Kline. The amplifier was loud, and she was off key. The blind organist's expression was fixed in a lunatic smile magnified behind a giant brandy glass containing a couple of dollar bills. The organ had a device that made rhythmic crashes like a hubcap full of marbles along with the melody.

Red neons around the bar bathed the clientele in a grotesque hue. Everybody was plastered. We sat on padded stools and ordered beers. At the end of the bar, a young man in a T-shirt tried to keep his forehead from dropping onto his beer glass. Next to him, three strong-looking women laughed loudly.

"Two of those broads are miners at Bethlehem," said Zurko. "Dikes, I think."

The blond bouffant finished her number with a flourish from the organ. The bartender led some scattered applause and the mike passed to another girl. Zurko asked what I thought of the place, and I told him it bored me.

"It's a slow night," he said.

We left soon after and picked up a couple of six-packs. Zurko always refers to six-packs as "traveling music." He has used the term even since we were teen-agers. We headed out of town and turned onto the back roads, which soon turned into rutted trails. He turned on the spotlight and flashed it through the woods, looking for deer. We were fairly drunk and into a discussion about life. I hadn't realized how bitter he was toward coal operators.

"They're pigs," he said. "They'll kill you in a minute if they can get a ton of coal out of it."

He took a beer off the seat and snapped the tab back. The beer squirted all over the windshield. He slammed on the brakes and I was thrown against the glove compartment. After much cursing and wiping with a rag from under the seat, we went on.

"It's so simple it's pathetic," he continued. "Miners are fuckin' miners from the day they're born. That's all they see. Their old man's a miner, their brother's a miner, they can step out of high school and make fourteen thousand a year, why should they make trouble? First it's the church says don't make waves, life's a fuckin' vale of tears, you get your slice after you're dead. Then the schools take over and baby-sit with the kids until they're ready to go into the pits. They can't even speak fuckin' English right when they get out after twelve years. The school makes goddam sure they can't do anything else."

We came to a Y that led to the tipple of mine #32 and we turned left through one of the huge pipes supporting the railroad trestle. We passed the portal and the parking lot and headed through a wooded section that would emerge on the Arcola road.

"You ever see southern Appalachia?" he said.

I said I had been through Appalachia a long time ago when I went to school in Virginia.

"You see what it looks like then. The land's been ruined forever, big gashes and shit soil. Now the operators claim they want to strip out west, and they say they'll reclaim the land. Can you believe anybody in the fuckin' world could look at Appalachia and believe those cocksuckers when they say it won't be the same out west? I used to love to fish that stream that runs by your cabin. Look what Nova Coal did to that fuckin' trout stream.

"You know who's behind it? The oil companies. How many people know that oil companies own two of the three biggest coal companies? You think they ever put any money into research for coal gasification or getting rid of the sulphur? Fuck, no. 'Cause they're making a killing on the oil. Those goddam Arabs couldn't do shit with their oil if they didn't have oil companies here distributing it for them."

I asked how he got to be such an authority on the country's economic situation, and he said from reading skin books like *Penthouse* and *Playboy*. We finished the beer and I got home about 4 A.M.

PART 5

February 3

COLD, ROADS CLEAR. DAY SHIFT, UP AT 5 A.M. I SAT IN the kitchen and watched the clock while bolting down a bowl of Cheerios and a banana and three cups of coffee. At 5:45 I went outside in the dark, put the lunch bucket in the car, scraped the windshield, shoveled the end of the driveway where the plow had piled up the snow. I put up the hood, took off the air cleaner, sprayed ether, started the car, put the air cleaner back on, and rumbled off to work.

The accident list: Jim Sprager, pipeman, burned both feet by going over his boots in sulphur water to repair a pump.

Hagley, a small, pale man, came over and said we were buddies. We were going to the slope to shovel out the tracks. Muck had built up so that the cars were in danger of derailing. The temperature outside was six degrees according to the radio on the way in, and it would be well below zero at the slope in the wind.

We sat together in the mantrip. As we approached the slope, everyone ducked down to avoid the icicles hanging from the roof and the piercing wind. When Hagley and I jumped out of the car, the wind almost knocked us down. We ran into a manhole to wait until the mantrip left.

When we came back to the tracks, Hagley picked out the muck along the rails and I shoveled it into a slit along the side where giant ice sculptures gleamed under my lamp. Each time we stood erect the wind drove us back a few steps. Our eyes streamed, hands and feet were numb.

Halfway through the shift, about 11 A.M., we went to a mechanics shanty around the curve to eat lunch. Yellow pumps and motors were stacked in the shanty and kept from freezing by a radiator of welded metal strips nipped onto an electric cable. Pumper, the garrulous old guy who takes care of the pumps, was hiding out in the shanty pretending to eat his lunch. Hagley slid the heavy iron doors closed and we sat on the pumps and opened our buckets. Pumper asked Hagley about his heart.

Three years ago Hagley was helping to haul 12,000-volt cable which was being pulled along a belt by a motorman. The motorman was several hundred feet away when a rear loop of cable caught on the belt structure. By the time Hagley could flag them to stop, the cable was stretched taut. He went to investigate. Just before he got there, he slipped and fell against the belt. The cable sprung loose like a slingshot. It cracked him across the head, threw him thirty feet down the shaft, and broke two vertebrae in his back. Three days later a clot in his head wound moved to his heart and he had a heart attack. Now, three years later, the doctors still could not induce his heart to keep an even rhythm.

"I'm making double payments on the house so the woman'll have a place," he said. "I'm only fifty-one. I'll never see that pension." Minimum retirement age is fifty-five.

The warmth of the shanty gradually thawed us. We ate our lunch slowly, savoring the heat. When we finally went back it was colder and the wind cut like a razor. I concentrated on not feeling and on the knowledge that the shift would end as all shifts did, a hot shower and drive home with the heater on full.

Chuck, the motorman, came by with a load of supplies and stopped to chat. He asked Hagley about his heart and said maybe Hagley could get a transplant.

"Prob'ly some nigger'd die and they'd put his heart in you," said Chuck amiably.

Hagley considered it. "No," he said thoughtfully, "I wouldn't want no nigger's heart."

"Well, you could run like a bastard," said Chuck, lifting his hand in a small wave and moving down the shaft.

After the shift, while we were waiting in line for the cage, we were standing on the ramp under some obscene chalk drawings. Somebody said that the drawings should be taken off before the women came in. Nick, an old boss, said, "You guys better watch your fuckin' mouths when them cunt come in."

Pumper said, "If they want to be coal miners, they have to learn to swear and chew snuff."

In the showers, I felt very light-headed. Driving home was like a dream. I took my temperature later and found that I had the flu.

February 19

TWO WEEKS AGO I CAUGHT THE FLU, AND I'VE BEEN HOME since then. Now the doctor says I've developed bronchitis too.

I have sinusitis and a migraine over the right eye. In the beginning I thought it would be relaxing to loaf around the trailer, but it has been impossible to sleep. An almost constant fever has burned ten pounds off me. I'm down to 178, and Grada says I look like an old man, hollow-cheeked, haggard. I am taking four aspirins an hour along with antibiotics. As I sit here in the trailer bedroom looking out the window, on the opposite hill, in a yellow, stubbled pasture a flock of crows hop behind a fence under a strange, luminous El Greco sky.

The past week has been constant rain instead of snow. From this window I watch coal trucks roar by on the highway below, spraying grime and black sludge over passing cars. The kids have been inside for several days because of the weather. How does Grada stand it? The baby has a cold from the wind that sweeps across the floor. The windows don't seal tight, even though they are framed with heavy duct tape.

I am brooding over the bills. It costs over $50 a month for oil heat in this tinderbox. With rent, telephone, and utilities, the monthly total is around $230. Food is another $50 a week. Grada says food prices here are much higher than in New York City. That's what she and her friends discuss in their letters. It seems impossible to explain the despair I feel now. Gas, clothes for the kids, mine clothes; if we ever get a few dollars ahead there is a walkout, an unauthorized strike, for three or four days and the next check is short.

This morning's paper said the Northland Steel Corp., owners of Nova and 18-D, increased their profits by 195.9 percent in 1974 over 1973. Northland is itself owned by some conglomerate. Meanwhile men like Clive who line their coffers for them lie on couches and wheeze for each breath in some tinny trailer.

This morning I told the doctor that I had to go back to work. I am not paid for days off. He nodded and said to put the pills in my lunch bucket. Now my bucket will be like everyone else's. On my way out of his office, the doctor added that sinusitis was usually much more painful underground because of pressure changes.

February 24

BACK TO WORK AT 5 A.M. COLD AND RAINY. NO TIME FOR breakfast. I ate a corned beef sandwich on the way. The road is usually foggy in patches where it dips through the woods, but today it was like driving through gauze from one end to the other.

I took the doctor's excuse, required for two or more consecutive days off, in to Kurtz like a school kid going to the principal's office. He wouldn't look at me and acted very annoyed, as though I had played hookey. He put on his ex-cop face and tossed the excuse to Cooper, who picked it up and peered at it. He turned the excuse over and examined it the way a cave man might appraise a cigarette lighter. His short, chubby frame, his habit of stroking his mustache thoughtfully, and his bewildered expression always suggest some distasteful perplexity, as though he can't imagine how all these fuck-ups came under his control.

I knew when he picked up the excuse that it would not be an ideal day for me. Finally, after reading it for several minutes, he said, "Go with Harry on the long wall preparation crew."

Harry was working with Richard, the young porno aspirant whom I had dragged pipe with during hoot owl.

Harry is a good man to work with. He is around fifty, white crew cut, and built like Tarzan's weight-lifting instructor. He had a tough job, moving cartloads of belt structure through low shafts, water holes, and muck, and gobbing rock (throwing it out of the road). I asked Richard where we were going. He just shook his head, and Harry said, "Goddam pneumonia parlor."

We walked away back into B-14, a low mud hole where icy water streamed from the roof and the muck pulled at our feet. Planks supporting the roof were bowed and cracked. Roof bolts hung out ready to tear chunks of flesh from our backs. The height was between a yard and 42 inches. I was dizzy and winded from walking back. Everything was covered with brown, dripping fungi. We loaded cart after cart full of awkward rollers and belt structure, then pulled it through the thick, cold mud and water, scraping our backs on the planks and roof bolts. When the steel three-wheeled cart was mired in muck, lifting it shook the pieces of structure stacked in it, and the pieces always seemed to jam against a finger. Our knee pads were quickly soaked and flopped down to our boot tops. To strap them tighter would have meant cutting the circulation. Pushing the cart from behind, I couldn't keep my head up to see where we were going, and I constantly smashed my head against rails and broken planks.

At lunch, Harry said, "I can't believe I've been in the goddam mines for thirty-three years. It seems like yesterday that I got out of the service."

He said he had been to my grandfather several times when he was a kid for various injuries and later had been a patient of my father's. He didn't have to ask what I was doing underground. As with most of the men, I was there because, for whatever reason, I couldn't be anywhere else. Richard said that he still planned to marry in June, but he hadn't found a job out of the mines yet. He spoke of his approaching marriage with a striking lack of enthusiasm.

During the last hour of the shift, Cooper came by and told me to move two six-inch cables to the side of the shaft. They were taut from the power box to the machine and wouldn't budge. "What's the matter, buddy, they welded to the ground?" said Cooper in his high, snide voice.

"Let it go," said Harry. "The bastard's not worth your time."

If Cooper heard him, he gave no indication of it. He moved on up the shaft and we started out the other way toward the cage.

When we reached the cage, we were almost at the end of the line. The two women were there, being bouncy, and Helen, thirty-six, the taller of the two, sang country songs between conversations. She has a pleasant, heart-shaped face with large blue eyes, a wide nose, and high cheekbones. The black hair curling around the bottom of her helmet is no longer than a lot of the men's, including mine. Pam, nineteen, is short and thin. She could pass for a boy except for her large breasts. Harry said she looked like a dresser with the top drawer out. Richard has been screwing Pam for some time, according to the razzing that he receives in the washhouse, but he didn't speak to her while we hunkered beside our buckets on the ramp. The men who did talk to them were smiling incredulously.

Back in the trailer, I went to bed for an hour, then sat around like a zombie. I sat in a chair with the kids in my lap, and wondered if I would make it tomorrow. In the evening, rain drummed on the roof. Grada sat on the floor doing a jigsaw puzzle and the dog slept on the couch. Grada noted that we are not much better off financially than when I started underground. Strikes, sickness, and inflation.

February 25

Snow covered the fields and roads. Halfway up a hill outside of River Gap, the car's back end went for the edge, a drop of several hundred feet. I put it into two semicircles, vastly overcorrecting to stay away from the edge, and it pulled itself out. Alone in the country in the dead of winter at 6 A.M., heart thumping with three cups of coffee, radio crackling, trying to drive as fast as possible without spinning off the road or bending a tie rod in a pothole, bringing total concentration to bear at the moment a coal truck barrels by in the opposite direction, all of this simply to feed, house, and clothe a family seemed suddenly ridiculous. I burst into laughter.

On to the mine. After suiting up, I was told that Crisco wanted to see me. Crisco, Cooper, and Kurtz are the mine hierarchy. Although Crisco is just a section boss, he is first among the other section bosses. After day shift, he, Cooper, and Kurtz gather in Kurtz's office and make top-level decisions. Crisco despises Cooper as much as everyone else does, but he tempers his views with an eye on Cooper's job as foreman. Crisco is about forty, with large, strong hands and a pot belly. His most striking feature is slightly bulged eyes, always open beyond mere alertness. Probably goiter. He has a pugnacious, assertive temper, like most bosses.

While I was waiting outside the office for Crisco, Billy

Sweet was waiting, too, spitting into the garbage can beside the water fountain. I asked him where he was going, and he said don't even think about it, you're going with me.

Crisco sent me to the office to pick up a wet suit. Billy already wore the yellow jacket and trousers. He said that if it was as bad as yesterday down there, he would walk out. Hank Kurtz, the super's nephew, was with us. We were all going with George, the timberman.

George had a run-in with Cooper over a year ago, a minor incident until Cooper insulted him. George blew up and wanted to punch Cooper. He followed Cooper down the shaft, calling him names. Cooper was humiliated and has never forgiven George. From that day, George has been sent to the sloppiest, coldest places, usually alone and under bad roof, that Cooper can dream up.

Billy Sweet is a hefty man. On the Legion scales one night we put a penny in for him and he weighed 233 pounds. He is about five feet ten inches tall. Moving that amount of bulk through a shaft with low roof is no cinch. Billy bid on and was awarded a supply job with Joe Morgan, a job that involves loading material and taking it to the face with a kersey. Then the women came in. Billy was taken off the supply job and replaced by one of the women. He could have gone to a committeeman right off and there might have been a strike, but Cooper and Kurtz knew with whom they were dealing. Billy hates to make waves. Committeemen are elected by the men in each mine to represent the men's interests and protect them from contract violations by the company. Our committeemen are young and aggressive, and would probably have liked nothing better than to take on Kurtz and Cooper over the issue, but Billy, like a lot of men, would rather work in slop than sit in an office with management and labor representatives and discuss grievance issues.

Billy said the slop hole we were headed for was full of strong sulphur water. After he had gone home yesterday, he said, the skin had peeled off his feet and his face had burned.

We all sat in the mantrip together. When we reached the slope, we climbed out and put on our rubber suits, taping the bottoms around our boots. Then we headed for the trapdoor. I held the door while Billy squirmed through, using both hands to force his knee pads over the sill. The wind tore off his helmet and threw specks of sulphur water through the door. After much cursing, we all crawled through the door to the low, cold shaft on the other side. The wind was a strong, constant irritant and the sulphur water gave off a noxious stench. The roof was low and cracked, the props bright orange and rotted. We went back seven holes, seven hundred feet, through water and muck that sucked at our boots with every step. Once the muck pulled a wedge out of Billy's hand and threw him off balance so that he went down on one knee in the water.

Hank and Billy voted to haul the props. George and I would set them, which involved digging for solid bottom while kneeling in a foot of ice-cold water. We had to pick a hole and shovel it clean without being able to see below the water.

Billy and Hank moved the cart with sixteen props a load through the water. They were soaked after the first trip, shivering, and threatened to walk out. Pushing the cart through water and mud was quickly exhausting work. They didn't really want to walk out because they had left an hour early yesterday. Besides, we all knew that George had been in this slop for a year and he had never walked out.

George said he had been working with Hagley a few days ago when Hagley complained of chest pains. He couldn't get his breath. George thought Hagley might have been having a heart attack, so he took Hagley out and drove him to the doctor. George and Hagley ride to work together and are good friends. After the doctor had examined Hagley and told him to take a few days off, George drove him home. He said Hagley's wife raised hell because Hagley took the following day off, even though it was doctor's orders. George said Kurtz had docked both of them a half hour off their sign-out time.

Hank and Billy finally decided to stick it out. It was a long, hard shift for all of us. Afterwards, in the wash-house, they said a man had been hurt up by one of the long walls. He and another man had been changing one of the big batteries in a kersey when the battery fell. He jumped back. The battery missed him, but he may have broken his hip when he slammed against another machine.

Another young man was hurt, this one at Harken mines. He was throwing crib blocks when one hit a prop, knocking it out and causing a rock eight feet by eighteen feet to fall on him. Five men were required to lift the rock. It had smashed the kid's back and ribs, but didn't kill him.

February 26

WIND SHOOK THE TRAILER SO VIOLENTLY THAT IT WOKE me at 3:30 A.M. I set the alarm fifteen minutes early so that I would have time to shovel the snow that had drifted over the driveway. When I was leaving at 5:30, the trailer door wouldn't close. The jamb was swollen from the snow.

Inside the washhouse, it was back into the rubber suit. Then we were off to B-14 again, this time with Hank, Billy, George, and Hagley. Everything was the same, except that I helped Hank and Billy pull the cart because Hagley was better than I at setting props. We decided that with three of us we could take bigger loads. We loaded twenty props a trip, so high that they sometimes caught on the roof. When I pushed the cart, I couldn't look up. It required total, exhausting effort to shove the cart through boot-level water for seven hundred feet. Once, I cracked my head against an overhead rail and it dumped me on my ass in the water. We had to stop periodically while Hank threw up. The sulphur fumes upset his stomach.

Hagley was very pale. He told us that last week he had been working alone, changing a prop under a rail. He was digging down to get solid bottom so that he could put a jack under the rail. While he was digging, one end of the rail suddenly fell and caught him on the neck. He was stooped just under the rail, so that it only drove him to the ground. He said if there had been any space between his back and the rail, the rail would have broken his neck. As it was, he had some bad bruises and several sprained muscles.

We had to push the cart down a short hill and through a water hole at the bottom. The muck was thick and water poured from the roof. We tried to gain momentum down the grade so that we could get through the pool without stopping. I was pushing when the hood on my slicker caught on an overhead rock and jerked me flat on my back in the water.

There was only one dry spot in which to sit or kneel in the entire shaft. After a full shift of wrestling with the cart and setting props, everybody was exhausted and wet. We came out fifteen minutes early. In the main shaft, we were dragging a fifteen-foot plank when Hank backed under a six-inch pipe gushing water from a pump. The ice-cold water sloshed down his back all the way to his boots. He jumped around the shaft like a frog.

Washhouse talk said that the kid from Harken mines won't walk again. Crushed pelvis.

Tonight in the trailer, Hadley sat in my lap and said, "Daddy, I don't have anybody to play with. I want to live in a house with a big front porch so I can play outside when it rains."

Just after she went to bed I received a call from the local dog catcher. He was giving me a $21 ticket because our dog had no current license. Our neighbor, Cutler, in the next trailer below, had turned me in. He called one night about a week ago, somewhat hysterical, claiming that Baron had backed his wife into a shed. I explained to him that the dog had never threatened anyone without cause, and, in fact, played with all the kids in the camp and took all the abuse they could give him. Cutler and his wife were not to be soothed, however, and the discussion ended with him threatening to shoot the dog, and me threatening to shoot him. I had not spoken to him before this incident.

February 27

THE ROADS WERE CLEAR UNDER THOUSANDS OF TWINKLING stars. I got into the parking lot early and stayed in the car for ten minutes listening to the radio, checking on what the rest of the world was doing. Vehicles careened into the lot and men piled out, lunch buckets shining under the arc lights. They moved toward the washhouse with their heads down, not talking, moving fast.

Inside, it was the same old thing. Crew bosses looked around the partition for their men while we suited up in the heat and dust. Tinny country music rattled over the speakers. Snuff or chewing gum in every mouth. The chatter while the baskets were down, swinging on the long chains from the ceiling, concerned who missed work, who had the easiest job.

Cooper, the foreman, sent me with Lou Perski. Some of the men refuse to work for Lou, but having been here only six months, I couldn't refuse. During a lunchtime discussion, one of Lou's neighbors said that Perski regularly beats his wife during their arguments and once threw her naked out of the house. The company fired him last year for being reckless with his men. He took the case to court and won, and the company had to pay him six months' back wages.

Three of us picked up wet suits. The two other kids, both eighteen, had just started last week. Lou took us down to shovel one of the main belts which runs deep into the mine. Since I was senior man, I was put on the wire side where cables hang alongside the belt. There was about two feet of clearance between the belt and the rib. The other side of the belt, the clearance side, had four feet of open space for passage.

Shoveling wet muck is different from shoveling coal;

the muck is heavy and sticks to the shovel. The belt was only about eighteen inches from the roof in some places. The muck had to be banged off the shovels against the cables so that the muck would drop onto the belt. We were hunched over on our knees in four inches of icy slop, extending the shovel to arm's length because of the low roof. Lou moved ahead of us along the main shaft. He snuck in through the trapdoors at the crosscuts and waited with his light off to make sure we didn't have a minute's rest.

We were moving against the belt, the wrong way to shovel. If the belt hit the shovel, the shovel would slam back at us. Meanwhile, my helmet and lamp cord continually tangled in the wires and my knee pads filled inside with wet coal that sanded my knees raw. It was hard to see. The lens on the lamp was always covered with slop. We did not get off our knees except for the half-hour lunch the entire shift.

When we came out, the next shift was in the hall waiting to go down. As we came out of the cage, covered from helmet to boots with several inches of gook, they started yelling, "Hey, what is this? Where's the bosses? Look at these guys!" No one appeared from the office to see what the uproar was about.

We went straight into the showers with our wet suits on and rinsed off as much of the crap as we could. Our boots would be wet inside tomorrow.

When the rest of the shift came into the showers, hawking and spitting black mucus on the floor and pouring bottled detergent over their heads, the talk was about going on strike over the women. There is great resentment, mainly because the women make the same wages we do, yet they can't do the work.

I stopped to visit Mrs. Hart after work. She was six

years old when her father was killed by a rock fall in 1908. He was thirty-four. She said that the last morning before he went to work he came back to the door four times to kiss her and her two sisters and tell them all to be good. He was nervous because he had to go into a shaft with unsupported roof. The company had told the men that props had been ordered but had not arrived. Later that morning, her daddy and her uncle were crawling back through the shaft with some timbers that they had found outside when a rock fell on her father and killed him.

The other men brought him home and buried him. After they buried him, the men put his clothes and belongings in a shed behind her house. She remembered going out to the shed with her mother a few days later. Her mother opened the shed door and fainted. A piece of her father's head and and parts of his body were still stuck to the clothes. She said that her aunt came up later and buried everything.

She and her two sisters were given away because there had been no insurance and her mother could not afford to keep them. She was given to a family of drunks that beat her and made her work on a farm from sun up to dusk. She ran away at thirteen, married, and had a relatively happy life. Now she had arthritis of the spine and constant pain. She was going blind soon, she said, but she kept going and trusted in the Lord.

February 28

LAST NIGHT I PUT VASELINE ON MY KNEES AND KNUCKLES. The skin had cracked. My raw knees gave off heat from the irritation. I was back at the same place today. One of the other men didn't come in. It was freezing, and the viscous muck was gummy from the cold. When we arrived at the belt, we had to wade through a knee-deep water hole. Then we had to crawl under the belt through the muck. We worked steadily, freezing, covered with sticky crap, our boots full of water, and finished the job an hour early. But instead of letting us go out early, Perski took us down to another belt. He was whistling, enjoying our misery. I would have loved to take him out in the parking lot and kick the shit out of him, but he is built like a truck. One of the few men I've met with an advancing hairline.

Near the end of the shift, I narrowly missed being hit in the face with the shovel when the shovel blade touched the belt. My feet burned, probably from a dam that we had gone through. The muck had been so strong that it took about a minute to pull each foot out. As we were leaving at the end of the shift, Perski said, "Cheer up, boys. You'll be back here tomorrow in the same place."

After we came up, I threw the rubber suit in the trash can. There was no hot water in the showers. The boilers were broken. I took a quick cold shower and dressed hurriedly. Some men went home without showering.

Going home, clear pale sunshine lit the countryside. At the trailer, on the desk-dresser in our bedroom, lay a pile of bills. Electricity is expensive because we have to keep an electric heater in the kids' room at night. Oil heat is expensive because the trailer's shell leaks like a sieve. When we were suiting up this morning, Knopick and a few other volunteer firemen were discussing a trailer that had burned last night. They said it had incinerated in five minutes.

I called the dog catcher and asked nicely if there wasn't

something we could do about this $21 fine. I told him it was half a day's work. He said I was lucky, the fine could have been $300. For an unlicensed dog? Sure, he said, serious offense. Then I called the magistrate, an elderly farmer who has been in office since I was a child. I was once in his house with a couple of state cops for a hearing about a ticket I was appealing. He and the cops had talked about how things were going, asked about each other, and finally we got to the proceeding. It took him two minutes to find me guilty.

He was very understanding when I called, said he had the same trouble with miners all the time. I could pay half the fine this payday and the other half next payday, he said. His fee was $11. The actual fine was only $10.

March 4

WE WERE OUT ON STRIKE YESTERDAY. I WENT TO THE union meeting last night at 7 P.M. through snow flurries and gusts almost strong enough to blow the car into small skids. The parking lot outside the union hall was full.

Inside, the local president, John Yunko, flanked on the stage by two other local officers, pounded the gavel and shouted, "This special meeting bein' now truly convened and now truly called to order. Raise your hand if you want to talk and the chair will recognize you."

Tables around the dance floor slowly filled with men ambling in from the bar. Blue and pink crepe flowers and streamers from a recent wedding reception hung from the girders overhead.

John explained that the walk-out was over a suspension notice given to a committeeman at 18-B, the mine next to ours which shares our tipple. The committeeman, a wiry young man full of nervous anger, rose and said he had simply been doing his job. He had stood on a bench in the washhouse and explained to the men what was going on in the office regarding safety violation reports submitted to the company. They had not been responding. The committeeman said he had wanted the men to know the company's position, the company being personified by the superintendent.

An informant had told the superintendent about the washhouse meeting. The super had not been on that shift, but based on what he was told, the super immediately suspended the committeeman for two weeks with the intention of firing him for inciting a strike.

John Yunko said that he and the other local officers had spent the afternoon with management, the super, and his superiors from the main office, and they had succeeded in getting the man's job back.

"That's the main issue. There's nobody takes bein' fired lightly, and when we save one of our brothers' jobs, it means something," said John.

John wanted everyone to go back to work on night shift. A group of men with the committeeman sat at a table to the left of the stage. They didn't want to go back until the safety issues were resolved. They felt that threatening to fire the committeeman had been a ploy to divert attention from the safety violations.

But none of us were being paid while out on strike, John noted, and there wasn't much solidarity these days. Men at the other three Nova mines were not happy about being out. It was an old story, John said. The company constantly bombarded the public about wildcat strikes. The newspapers had neither the space nor the will to explain grievance procedures to the public. The coal operators, with unlimited funds and legal talent, exerted a constant pressure on the contract, often breaking it just to provoke a test case. They forced hearings, winning some, losing some, always keeping the union on the defensive. When miners walked out, people only read about another wildcat strike and shook their heads over those asshole coal miners pulling that strike crap right in the middle of our energy crisis.

A young man at the table with the committeeman stood and insisted that getting the job back wasn't enough, that the issues about seniority and safety which had caused the trouble in the first place were still unresolved.

There were about two hundred men in the room. As he spoke, he looked around at the group and lost control of what he was saying. But he didn't want to sit down feeling like an ass, so he berated the operators for another five minutes in a rambling tirade. Then someone else took up several of his complaints and enlarged on them.

After a while, Bellinski, burly and fortyish, wandered in from the bar. "Mr. Chairman," he shouted, interrupting the kid in front. "Goddam it, are we going back to work or ain't we? I don't wanna hear about fuckin' buggy runners' seniority. Are we goin' back or not?"

John massaged his temples and looked down at the table. "Jesus Christ, Bellinski, that's what we been talking about for an hour. Go on back to the bar."

The kid in front was somewhat insulted by Bellinski's remarks, especially since Bellinski had voiced the view of the entire rear section where I was sitting. Men shouted agreement. John Yunko told the kid in front to relax.

"Listen, the issue is settled," said John. "The man has his job back. Now, when we went out there this afternoon, we wanted to ask whoever the person was who told the superintendent about our committeeman in the washhouse, we wanted to ask him some questions. Since the super got everything secondhand, you know. Well, they weren't very eager to produce this person, and I hope it was a boss and not a union man, because none of us would think very much of the sonofabitch of a person. They wouldn't produce him. We got the man's job back and the other issues will be handled. So it's up to you men. You got to decide if we go back or if we don't."

The kid with the committeeman stood up and started another wrangle. Mitch, at the table with me, threw up his hands and said, "This is enough for me." He walked out and I followed.

It had been snowing while we were inside, a couple of inches of new, light stuff. I had to go about a quarter mile up the back road to turn around without danger of getting stuck. When I came back, someone was trotting out of the union hall. I asked him how the vote had gone and he said we were going back tomorrow day shift.

March 6

HOOT OWL. I SET TIMBERS WITH HANK KURTZ AND A short, stocky guy called Endive, five feet two, who can actually run in a crouch through the shaft in some places. Long, dirty dark hair and a beard make him look like an Irish dwarf. Hank is just out of high school. Endive, twenty-four, is married, has two children, and has a month more time underground than I. We were off in an escapeway along Main B, a warm, safe, and dry spot.

After hauling props from a rail car out in the main and stringing the props through the shaft, we sat and talked for the rest of the night.

Endive said he had been bumping his uglies with Pam, one of the girls who has just come underground, for years. He told Hank that he would take him down to her trailer next week. He said her fifteen-year-old neighbor gave great head. When asked how he was getting along with his girlfriend, Hank said he wasn't getting tit yet.

"That's the way you want it, buddy," said Endive. "If you could fuck her, what's to stop anybody else from gettin' it? Then you can't marry her, right?"

Endive said that he had a buddy, evidently in our mine, who snorted cocaine every lunch break, but I got the impression by the end of the shift that Endive's imagination puts in a lot of overtime.

At the end of the shift we were crawling through a crosscut when I heard a solid *whump!* behind me. Hank said, "Jesus, that was close." A pothole, a conical piece of rock about two feet in diameter and weighing about fifty pounds, had dropped out and narrowly missed his leg. Potholes, or niggerheads, come in sizes up to five feet or more across. They look like circles in the roof and drop without warning.

March 9

WE WERE BACK AT WORK AFTER STRIKING FOR TWO MORE days because the truckers were picketing. Last night we set timber again along the warm, safe, and dry airshaft off Main B. I was working with Hank and Helen, the blue-eyed friendly brunette who hums country songs. She sat on the edge of the three-wheel cart and regaled us with her life while we set props. She had to stay up-wind because Hank was hung over and farting. We had to haul another rail carload of props through some low roof, a full night's work with the small cart. The low roof always opens old sores along the backbone, but it was the best that Hank or I had had it for a long time. Helen stood next to the rail car in the main shaft while we unloaded it, cautioning us not to pick up splinters.

She is thirty-six, with seven kids, ages thirteen on down and lives on a farm with her boyfriend, a buggy runner in another mine. He has five kids. Helen used to live in Oregon, where she said she'd been a Hell's Angels' mama. She promised to bring in pictures of the bikes and the young sports she rode with.

She said she was their banker. When they were busted by the police, she'd wire them money. They couldn't take it with them, she said, because they were always fined whatever amount they were carrying. She seemed delighted that Hank was only eighteen, and they talked about fucking for a while. "My favorite thing is when you fuck for so long and do it so good that when the guy gets up he can hardly walk. You know, like his legs shake. But I got to love a guy to fuck him. I just cain't get it on unless there's feelin'," she said. She wants to have a shift party on her farm, all the men on our shift. She said she wanted to come to work stoned sometime. I said I couldn't imagine using drugs underground, and Hank answered that a lot of men do. He mentioned Smilin' Jack on the long wall, a young fellow named for his constant grin. Hank said he was a speed freak, that was why he was always so happy.

March 11

SECOND SHIFT THIS WEEK, 3 P.M. TO 11 P.M. I WAS SENT to the end of Main B to set seven-foot timbers with a new kid named Vekar. He has been in the mines for a week. Asked how he liked it, he answered, "I like it. I really do." So far, he had been Clive's buddy and they had been high and dry. I suggested that he reserve opinion for a while.

He said he knew it wouldn't always be easy. He laughed and told me a story about one of his brothers-in-law who had been in the mines. The first day, he had been sent to the long wall. When he came home, he said he didn't know if he could handle it. The second day on the long wall, he suddenly started shaking. He covered his head with his hands and tried to crawl under things. They took him out and he never went underground again.

At twenty, Vekar is married with no kids. He has a mortgage. He said he used to work in a metal plant for $3.25 an hour, about half what he is making in the mines. He is a very nice, quiet kid with straight black hair, dark eyes, and a quick smile, who seems to want nothing more than he has.

The props which I had brought down in the rail car weren't long enough. We had to cut four-inch planks and carry them about fifty yards, then dig down to hard bottom, set the plank, and wedge it in. We set fifteen posts, not a bad night's work. I was tired. Vekar kept offering me coffee from his Thermos, helping the old man through the night.

We were a little late coming out because my watch was slow. We stood beside Helen in the line waiting for the cage. Last week's hoot owl shift had left deep lines in her face and her color had waned to an ashen pallor. The long wall crew arrived, shouting exuberantly at each other and horsing around. I wondered why they were so much happier than the other crews. Relief at being away from collapsing roof?

March 12

I WORKED AGAIN WITH VEKAR SETTING TIMBERS ALONG the escapeway parallel to the main shaft. We had to crawl under the belt and through a trapdoor, or escapeway, just as we had been doing for a week or so, but for some reason I couldn't remember which one we had used. Crosscuts are one hundred feet apart, and missing by a few doors means scuttling like a crab through the low instead of walking in the main shaft.

Actually, in most places we had to roll over the belt because there was not enough clearance underneath. This belt isn't ordinarily in operation between shifts. We couldn't shut it off because the jabco runs only on one side, and after we were across we couldn't reach it to turn the belt on again. So we rolled over the belt and hoped it wouldn't start up suddenly and take us on a fast ride out to the tipple.

We chose the wrong door and had to crawl through the low for a couple of hundred feet. Our work area was near a set of air regulator doors, sliding wooden doors that control the wind speed.

We worked steadily for a few hours setting props. I asked Vekar how many of his high-school class used dope. He said over half. Grass, acid, and downers mostly. The other half used booze. We were talking about how operator-miner relations would change in five years when miners were on dope instead of booze. Then one of us noticed the silence. We exhaled and watched our wisps of breath curl straight up instead of dissipating down the shaft.

"The fan is down," he said. "Do you remember how long we're supposed to wait?"

"No. It's either wait a half hour or we have a half hour to get out. You just finished that three-day school they send you to. Don't you remember?"

The fan kicked in for a moment and went off again.

"It's probably just something knocking the power off. They'll fix it," he said.

We went back to work. It was some time later when we both stopped at once and noticed the silence had returned. I hit a prop and we listened to the echo bounce through the shafts for a long time. I never realized how much noise there was underground until everything stopped. It was warmer; we were sweating.

"Did you see that?" he said.

"What?"

"I think I saw a rat running out the trapdoor."

I turned my light on the door and caught the white back end of another one disappearing over the sill.

I remembered what an old miner had once told me in a bar: When you see the rats leaving, grab your bucket and run.

We crawled out the trapdoor and under the belt into silence. No movement, no sound. We walked along the main shaft quickly, listening to the crackle of voices over the phones. After about five hundred feet we encountered the first pool of black water. When the power goes, the pumps stop, and the mine starts to flood. We walked along the rail and used the roof to balance ourselves. The next pool was too deep, and icy, brackish water poured into our boots. On the other side of the water, Vinny, a fat little mechanic, sat on an oil can and wiped his face with a rag. We were all sweating, but Vinny was bathed in it. He was waiting beside a phone, he said, to assess the situation. A boss came out of an escapeway, breathing hard, and told us to get moving. It was illegal to be in the mine with no air and we had about a mile to walk out.

We met other men on the way to the long, steep, forty-five-degree-angle slope at the far end of the mine. The slope provides ventilation and it is also where the coal goes out and supplies are brought in. By the time we reached the slope, Vinny and the older men with black lung were puffing and wheezing. Clusters of headlamps about every hundred feet up the slope marked groups of men resting in the stiff wind. The slope was covered with gritty, dangerous ice patches. If a man slipped, he could have a long, hard slide and pick up a back full of cinders.

Some of the older men must have been on the slope for a half hour or so trying to catch their breath. Walking out was a chore for the lungs and the cold wind turned the sweat clammy. When I came out at the top, emerging under the steel superstructure that supported the main belt, there were two trucks waiting. The big one, a thirty-two-ton that we had to scale like a barrier in an obstacle course, was for younger men. The other truck was a walk-in for the black-lungers.

The big truck filled and we took off for the washhouse, the engine roaring through the gears up a long hill past dark houses. The men shined their lamp beams through the windows and whistled. I wondered what the people inside in their beds must have thought. We passed fields streaked with snow and lined with grotesquely twisted dead trees poisoned by the sulphur from the nearby settling ponds, then went on through a sparse stretch of woods where our beams bounded through the trees. When we reached the washhouse, no one expected hot showers. The boilers went off along with everything else. But we were the first to arrive and the water was still hot, though the pressure was low. The older men arriving after us blundered around the hanging baskets, trying to undress themselves by the light of their headlamps, bitching when they learned that the hot water was gone.

When I went outside, the next shift was just arriving. They knew something was up because the tipple lights and the arc lights in the parking lot were out. They roared around the lot yelling, "What the fuck's going on?" When they learned that the fan was down, most of them spun around and headed home. A few parked and went inside to collect four hours' shape pay by sitting in the washhouse.

We were out early and the River Gap Hotel was full. I stopped with Vekar for a few beers. He was quiet until we had finished a few rounds. Then he said, "You know, I'm not sure I want to stay in the mines anymore. When things pick up I think I'll look around for something else."

We were next to John, a big, gray-haired track layer. John laughed so hard that he had to put his beer back on the bar. When he had composed himself, he said, "You do that. You look over your car payments and all the other crap and then read the want ads. You're married, right?"

Vekar nodded. "A year ago."

"Pretty soon you got a kid. The place ain't big enough you got now. The old lady wants more house. By then you got a year, maybe a couple years in. You know any jobs outside settin' timbers or runnin' a Lee Norse? Look at it this way. There's this long tunnel you're in, and it's gonna take you another twenty-nine years to get through it, crawlin' all the way."

Vekar didn't say anything. He left after that round. John watched him go out the door and said, "That'll give him something to think about."

When we finally closed the bar and went outside to the porch, a light snow was falling. I zipped my jacket while we clumped down the wooden stairs to our cars.

"John," I said, "if you saw your whole life the way you told it to Vekar, how could you face every day?"

He shrugged and thought a moment. Then he shrugged again and raised his hands. "I don't know," he answered. "What the hell else can you do?"

March 14

I CUT WORK TODAY. LATE THIS EVENING I WENT INTO THE Greenridge American Legion, across the street from the church. The Legion was paneled in pine and had indirect lighting of the bright, no-nonsense variety behind the bar. A few other men and I drank silently under the bright eye of the color TV until the dart league arrived.

The dart league was mostly miners, all of whom were delighted to be away from their wives. Whooping and hollering filled the bar, the pinball machine clanged. It was impossible not to be drunk after a while because rounds were bought in lots of eight or ten without regard to how many bottles were still untouched.

Zurko came in. We leaned against the long radiator between the bar and the men's room. He looked insane, bursting with energy, eyes wide and snapping. He was wearing three or four shirts—checked flannels, T-shirts—heavy trousers, and work shoes laced halfway up the calf. I told him I had cut work and he laughed.

"The time you make a full week, I'll buy you a case of beer," he said. "You're like the rest of these part-timers here, you work five days because you can't live on four."

Billy Sweet, the big fellow I had set up timbers with in the slop some time ago, came in with his wife. They sat on stools next to us. Bill's wife had delivered their fourth girl recently and we congratulated them. Zurko added that it was too bad that Bill couldn't make a boy.

"Don't worry about it, Zurko, we'll git one yet," Bill said. He turned to me. "Say, Buddy, we didn't see you out today. We missed you."

"Spent the day in the woods with the dog," I said.

"The glamour and romance of bein' underground is losing its charm for him," said Zurko.

"No," I said, "I just want to know how so many men can do this and not start a revolution."

"They're too fuckin' dumb," said Zurko.

Bill's wife laughed and Bill said in his deep drawl, "Now, wait a minute, Zurko, we ain't all been to college. You're down there too, you know."

"Nothing to do with college. They convince you early it's God's will to accept your lot. You get the pie in the sky later. Meanwhile, they don't even teach kids to read and write when they get out of twelve years of school. Kids can't speak fuckin' decent English. It don't make a goddam how smart you are. Rich people aren't smart, they just grow up rich."

"What do you know about rich people?" said Billy.

"I read the fuckin' papers," he answered.

Zurko comes from a large family, eight or ten kids. He was the only one to have gone to college, for all the good it did him. His father, a very nice, quiet man, put his time in underground and is now retired. Both of his parents live in a small company house that they bought and where Zurko grew up.

Later, I went with Zurko in his truck to another bar. Amid the cascade of violins from the speakers in the back, we sped through the dark while Zurko raved angrily about life in general.

"The goddam country is run by fat-assed crooks. They're heroes now, like pimps to nigger kids. Agnew, Nixon, they all got away with it, right? You ask people what they think and they fucking shrug their shoulders and say that's what it's all about, man, get what you can and get the hell out."

He was yelling over the speakers. I had a headache. We hit two more bars while he continued preaching about who poured the water softener in America's moral fabric. I couldn't wait to get home.

March 15

ANOTHER FOUR TO SIX INCHES OF SNOW FELL LAST NIGHT over the sleet. Talk in the washhouse was about Helen. She had gotten lost underground yesterday and it had taken her five hours to find her way out. Today, we were sitting on the bench when Superintendent Kurtz came over and gave Helen a winning smile. He told her he was putting her with Joe Morgan, the supply man whom Billy Sweet usually works with. "You keep him in sight at all times. When he stops, you stop. When he sits down, you sit down," said Kurtz.

Morgan, a big man and a hard, steady worker, was

resolutely against working with a woman. His wife had already made a few remarks about women and men working together underground. The main problem, though, was that he would have to do all the work himself, including loading 1,250-pound rails, planks, fifty-pound bags of rock dust, while she sang country songs to him.

I was back with Vekar in the air shaft, a safe, warm, and dry spot. A man could learn to like the mines if it stayed that way.

After the shift, while we were waiting for the cage, Meroni, the section boss, joined us. An old Italian, one of the nicer bosses, he couldn't figure out what in the world women were doing in a coal mine. He hadn't the vaguest idea of how to treat them, so he pretended that they didn't exist. We stood together quietly for a while. Then he bent over and whispered to us, "You know, I don't have to guts to do it yet, but one of these days I'm going to ask these women what they do down here when they get that monthly thing. You know? I mean, they got these pom-poms, right? But how the hell are they going to change them down here?"

Jerry, the belt cleaner, said, "Yeah, next you'll see a big rat runnin' around with one of them bloody things in his mouth."

Meroni laughed and laughed, shaking his head, but afterwards he was still looking at the women in the line behind us, wondering.

Pumper asked Pam if she'd give him some pussy down there, and she replied, "I'd be afraid to give you any, Pumper, you might have a heart attack."

A chain runs down the center line of the eighty-foot cement ramp from the elevator to the bottom doors where the ramp empties into the main shaft. The men waiting to go out were on one side, filling the elevator twenty at a time. When the cage came down again, it brought twenty men from the new shift. They shuffled down the other side of the ramp. Electric bulbs are spaced along the cinder block walls, so most men waiting to go out turn off their lamps and pull the light off the helmet, draping the cord over their shoulders. Some take the battery off the belt and wrap the lamp cord around it so that when they burst from the elevator at the top they can shove the battery into its charger, hang up their lamp check, a metal disk with a number on it, and charge into the washhouse for a shower. There are about forty showers for the whole shift, usually about a hundred men, and the competition is stiff.

In the hall at the top of the cage, the men wait to go down and watch the previous shift come through, faces black, eyes dull with fatigue. A few smiles, a few greetings among friends. The new shift at the top sit on benches with their backs against the corrugated tin walls. Some spit snuff into nearby buckets, others tap their hammers on the floor.

At the bottom of the cage where we were waiting, there is a level square just in front of the elevator where we hunkered or sat on our buckets, backs against the wall. Pam was back along the line telling about the crew she was working with. During lunch, Pam and the helper had rigged up a piece of canvas which hid them from the boss's view and also kept the freezing wind from blowing on them. Suspecting the worst, the boss ordered the helper to take down the canvas. The helper complied, and Pam had yelled at him to put it back up. "I told him all I wanted was to keep out of the goddam wind," she said. "I told the boss if I'd wanted to fuck him I would've taken him out in the parking lot when we get out of here."

Big John, the track layer, muttered behind me, "If they were decent women they wouldn't be in the mines. No decent woman would come down here."

March 17

ANOTHER MONDAY, DAY SHIFT. I WAS SENT DOWN TO A-17 with Billy Sweet and Hal, a twenty-year-old with about a year and a half underground. We were to come-along a tail piece out of the shaft, over a belt, and into the main heading.

The tail piece, the end of the belt drive, is a flanged hunk of steel that encloses a big roller. It probably weighs around a quarter ton. Come-alongs, ratchets with three-quarter-inch wire cable on a spool, have a hook at the end of the cable and another hook on the ratchet end. The come-along is hooked onto something solid, such as the plate of a roof bolt, and the other end of the cable is hooked to what is being hauled. Then the pipe handle is cranked, winding in the cable. A ratchet gear holds it after each wind. That is the idea, at any rate. But while we pulled the come-along through muck and water, the cable snarled, the ratchet gummed up and slipped, and sometimes the cable snapped and whipped back like a snake.

We hooked the come-along onto a rail and fastened it to the tail piece. After a few slips, the tail began to plow through the mud. Then Cooper, the foreman, arrived. When Cooper appears anywhere in the mine, conversation stops and everyone stiffens, like a pack of dogs ready to fight. Cooper sat there for a moment, flashing his light over each of us, and said, "They sent three men down here to come-along a tail piece?"

We had been bitching because we should have had four men. Cooper thought that there should have been only two of us. After a few snide remarks on the quality of the help, he left and we resumed pulling the tail piece out.

At lunch, Hal was talking about his budget. He has no children. His wife is a high-school senior. "People think miners make a lot of money, but at the end of the month I got car payments, rent, bank payments, jeep payments. I got nothin' left."

March 18

A FAKE WHIFF OF SPRING. A FEW ROBINS, WET GROUND, soft breezes scented with pine and humus. It was warmer underground, but instead of woodland aromas we were treated to the stench of rotten eggs when we pulled out an old prop.

Today I went with Aldo, a brattice builder and a tough, surly whiskey drinker, a fighter whom the bosses avoid riling. Even Cooper won't mess with him. Aldo is an iron freak along with handling cinder blocks all day. His hands bulge with muscles and tendons and his forearms are thick and marbled with blue veins. He is thirty-one, with six years in 18-D. That gives him considerable seniority, since the mine is only eight years old.

We had several hundred blocks to bounce. We crouched with our backs against the roof and tossed the blocks down the shaft, crawled down and tossed them again, and finally threw them through a trapdoor into the next shaft. Aldo bitched a lot in a deep, raspy voice. He told me some stories about his chopper days in Viet Nam as a tail gunner with a .60-caliber machine gun. He said their crew chief was crazy. He used to order them to rocket the hooches. "Women, kids, we didn't know what was in them. But you can't trust none of them. They shoulda pinned a medal on Calley's chest."

He said that after a while the crew never returned without the chopper being riddled with bullet holes, and once a round missed Aldo's head by inches. He decided the hell with it then, he said. They made a sling and took turns hanging from the chopper, twisting around and raking the ground with machine gun fire. The crew chief and another gunner were shot, he said.

Last year, Aldo was struck down in a hit-run accident and his shoulder was broken. The doctor guaranteed him arthritis if he stays underground in the cold dampness, but Aldo is getting married in June and wanted to keep his seniority until a tipple job opened up. The shoulder pains him often, but he is proud of what he has done with it.

"When I came out of the hospital, the doctor said I'd never have full use of my arm again. I couldn't raise it more than halfway. I started my own program, got out my weights. Now look," he said, raising his arm over his head. "If I get arthritis, I get arthritis. What the fuck am I gonna do, right?"

We built two walls. Aldo plastered half of one for me to make it airtight, since I didn't have the knack of laying on a thin, solid sheet.

In the mantrip, everyone seemed happy. Probably because of the warmer weather. Talk of softball, fishing.

Pam, the short, big-breasted girl, was sent to the long wall today. Water runs from the roof in torrents up there now. Everybody was pleased about that.

In the mantrip, Endive, the dwarf who said he was taking Hank down to see Pam, told us about screwing her on his bike one afternoon. "I told her I always wanted to take a piece of ass on my bike, so she got up and leaned back over the handlebars. She wanted to get on the ground, she goes, come on, we can move better down there and all that, ya know, but I go, no, I had to get one on the bike. Then after we got on the ground, I told her now we can't get on the bike again 'til the stains dry on the seat."

Endive said he was crazy about his two small kids. To entertain them, when he arrived home from work everybody would pile into the car and they would go for a ride, sometimes eighty miles or more.

After work, the sun was shining and the air was warm and filled with bird songs. Splashing groups of mallards were fooling around where several ponds merged at the end of the mine's road.

I stopped at the River Gap Hotel for a beer and met Zurko. He had taken the day off to fix his truck. He had just had a haircut, and his cropped black, curly hair and gray sideburns framed the tension lines across his forehead and around his eyes. Though it was cold in the bar, he wore only a torn gray T-shirt and jeans. He might have been drunk, but his impatient manner and undirected energy made it hard to tell. Often, he seems drunk when he isn't and sober when he is completely plastered.

He rambled about a place where men were sentenced to life at hard labor and a lingering, painful death. They were treated with contempt, and if they revolted, they and their families were left to starve.

"That's us! You see it?" he exclaimed, thumping his fist on the bar.

I told him I thought he was cracking up. He admitted that it was true. At thirty-eight, he saw no way out. He said that lately he was taking an ice bag to bed with him to put his feet on so that when Judy, his wife, rolled near him he could plant his feet on her and send her away. He was too much for me and I left after one beer.

March 19

BACK WITH ALDO BUILDING BRATTICES IN THE MAIN AIR return. It was important that the walls be airtight, so I stacked the blocks and helped him build the walls, then Aldo plastered them while I sat around. He talked about his fiancée.

"This ole gal, she ain't very big at all, but she moves all the time. Always gets around. She keeps her apartment real clean, that was a thing I noticed. I'm particular about that. When I get out of this shithole every day I want a clean place to relax in," he said.

He was plastering, throwing the cement on and smoothing it over with long sweeps while he talked in a loud, graveled voice. Aldo keeps tranquilizers in his bucket to help hold his temper in check. Even with that, the inevitable foul-ups underground always put him in a bellowing rage for the first hour we are there.

"She straightened me up, that little woman did," he said. "Christ, I was out gettin' fuckin' drunk every night, fightin' all the time, raisin' hell. I dunno, I guess when you get older you just get more brains. I'm thirty-one, I guess it's about time I got settled. Anyhow, I'm on probation for two years, I can't afford no trouble now. Remember the Sykes Company tipple fire? Well, I was one of the eight guys the cops nailed for it. I wasn't even near the fuckin' thing. I was down on the railroad tracks talkin' to a state cop when it went up. The district attorney's office wanted to make an example of us. It was a scab mine, you know."

He finished the wall, scraped the bucket clean, and came over and sat on a pile of blocks. Once the wall was up, it strengthened the air flow and forced a stiff wind through our shaft. I was stacking the blocks into a wind break while Aldo poured a cup of coffee from his metal Thermos and sat down.

"They wanted to organize it, the union did," he continued, "so they pulled everybody out in District Forty-one. We all went down to Sykes', five thousand fuckin' miners, and the union brought in a couple trucks of booze to get the men stirred up. The railroad tracks were here, in front [drawing a line with an ax], and here was a couple hundred state troopers shoulder to shoulder. Behind that they had the washhouse and the tipple. Sykes' men were up there with rifles, so we went to get the fuckers. They ran out the back, down the hill in a truck. I went back out front, I was talkin' to a cop, askin' if any of them knew my brother-in-law. He's a state cop. They said sure, they knew the fucker. They wanted us to burn the fuckin' tipple, they kept sayin' why don't you burn the sonofabitch so we can go home. This one corporal says, 'My old man was a miner. I know what you're talkin' about. Burn the sonofabitch,' he says.

"Strikers tried to burn it the night before, there was a big boney pile in front and they were shootin' arrows with gas soaked rags around them, but it wouldn't burn. Well, after a while a lot of men got liquored up pretty good and some of them went around back. There were only a few cops back there. They rushed the cops and knocked them outta the way, then they threw them cocktails in there. That baby burned then, Christ, you could see it for miles.

"I got into it with a cop. There was three of them beatin' the shit out of a guy with clubs. They had his head laid open and I told this one cocksucker, I says you call yourself a man, take that fuckin' gun off and we'll go down around the hill. Then we'll see how much of a fuckin' man you are. He says buddy, if I could I'd be glad to, but I can't. That fucker testified about it at the trial, too.

"At first they arrested eighty-nine, then they let twenty go. Then they just charged the eight of us. My brother-in-law said they wanted to make examples of us. They took one guy from Tippleside, one from Arcola, and Pillartown. They got us for unlawful assembly, then my brother-in-law said they wanted to charge us with riot. They had pictures of everybody. Our lawyer said if we would identify the men in the pictures we'd get off easier, but I told them to go fuck themselves. They said they wouldn't prosecute those guys, but I says what the fuck you want their names for then. Fuck those cocksuckers, I says, let them do their own dirty work. They wanted to make a deal, they says we'll drop the riot charges if you identify the other guys in the pictures, but we said fuck you. Then they said if we pleaded guilty to unlawful assembly they'd drop the riot charge on all but three guys. We said we'd go to trial together. Tony, one guy, he had a nervous breakdown and killed hisself. He drank a fifth straight down. His buddy was with him. He was so drunk, his buddy didn't have time to take him home so he took him to work with him. He went out to check on him about ten o'clock and Tony was dead."

He shook his head. The wall, already drying in white patches at the end where he'd started, was finished. The plaster was tight across the roof, the blocks were sealed and the sides tight. Aldo washed the bucket out and scraped it with the trowel while I stacked the few remaining blocks. We took a moment to play our lights over the wall. Building brattices is satisfying; at the end of the shift we could admire what we had accomplished. Tomorrow in the office they might change their mind and have it knocked down, which would set Aldo off again, but it was day-to-day living. Aldo finished his coffee and we went crouched through the shaft, carrying the tools, to the mantrip.

We were early and took the first seats on the rough plank that runs longways inside the car. Empty snuff cans and half-eaten sandwiches, Mrs. Smith's pies, black, gelid hawkers, mashed doughnuts covered the floor. We pulled our jacket collars up against the stiff wind. When conversation died, we silently played our lights over the broken rocks hanging between rails and the cracks that zigzagged across the roof. Some props were split, others dry-rotted and cracked, sparkling with fungus in orange and white clumps. The sides and roof were gray, having been machine-dusted with powdered limestone. Black patches appeared between the rails where rocks had dropped out since the last dusting trip.

March 20

CLOUDY, COLD, WINDY, SOME RAIN. I SPENT THE DAY cleaning out and timbering the shaft under the elevator. Dan Kurtz, the superintendent's son, and I set props and knocked out a door at one end. A nice, unassuming kid with sandy hair and long eyelashes, slight build, medium height, Dan had graduated from high school last year along with his cousin, Hank. His ambition was to work in the mines for a year, save some money, and become a truck driver.

The shaft was narrow and about a yard high. We shoveled the coal onto a piece of canvas and dragged the canvas about a hundred feet to the main shaft. Then we shoveled the coal off the canvas into an area behind some props. I asked Dan what high school was like these days. He said drugs were everywhere, whatever anyone wanted.

"We had some dynamite THC last week, and some great windowpane acid. I loved that, I saw gold dust dripping off my hands and everything. I think I'm doing too much acid now, though. I do it when it's around, and it's around every weekend."

He estimated that sixty to seventy percent of his class used dope. He said a kid that had had two hundred downers in his locker and they caught him. He was suspended for two days. Another kid took acid and thought he was Jesus. He shot up a church with a shotgun.

Dan had been on the long wall for a few months working for his older brother, Elwood, who bossed up there. Dan was missing too much work and his father had warned him that he'd take him off the long wall if his absenteeism continued. It continued, and Dan left the long wall. Dan still lived at home. Elwood, married with two kids, lived across the street. Dan said he would come home after a night's partying with barely time to change his clothes before going to work, and his brother or his brother-in-law, also on the long wall, would be waiting in the car to go.

"I don't need the money that bad," he said. "I just told my dad he couldn't threaten me, so I got took off the long wall. I really liked it up there, too."

A constant, cutting wind froze the black puddles around the edges. Our toes lost feeling and hands turned numb. After loading the coal, we went out to the main shaft and put on our jackets. It didn't help much. We were caught between working to stay warm or sitting still and shivering. We had to fight the tendency to loaf even when it was uncomfortable.

Dan said he had two friends who rented a trailer where I lived. Their trailer was probably the source of the rumbling bass notes that buzzed through the floorboards at night. The state police patrolled the camp regularly, Dan said, sniffing around for people to bust.

He asked where I had been working lately. I told him with Aldo and he laughed. "Whiskey drinker," he said. "Did he tell you about when he tore up that state cop? When they were burning the tipple? Man, they say he beat the piss out of that cop."

I said that Aldo hadn't mentioned that part. Dan said that that was the reason they had indicted him.

March 21

HARD RAIN. YESTERDAY ALDO HAD THOUGHT I WOULD BE working with him. He had spent the morning going up and down the shafts hollering, "Hey, Arble! Hey, Arble!" When the boss came by in the afternoon, he asked Aldo what he had accomplished. Aldo said, "Not a fuckin' thing. I been lookin' for Arble all day. I walked from the slope to the cage. You give me a man, you take him away and don't tell me shit. . . ."

Aldo had actually put up a wall or two. He just wanted to impress the boss that he hadn't had a helper and was expecting a man he couldn't find. The boss, a nice older man named Beldak, said he'd pass on to the office what Aldo told him.

Cooper stopped me in the shaft and said, "Why don't you be more careful in your work? You put a hole in a waterpipe yesterday down in Main B."

I told him I hadn't been in Main B yesterday. He walked away. Later, it turned out that Crisco had confused me with another man because we both had goatees.

I was back with Aldo. We had to bounce 250 cinder blocks for two holes (two hundred feet), then carry forty-five-pound bags of cement for the same distance through a shaft forty-two inches high, a strain on the lower back. We couldn't drag the bags because we would have torn them.

After work I went into the office and told Cooper that he had made a mistake—I hadn't broken the pipe. He was leaning back with his feet on the desk and stroking his little mustache. When I asked if Crisco had told him about his mistake, he nodded and said, "Yeah, he told me." No apology, no nothing. As Dan said of him, "Cooper's the kind of guy, if he ever dropped acid he'd see a boney pile."

Aldo said Cooper never speaks to him. When Aldo had something on his mind, he would find Cooper and yell at him for a while.

"He just looks down," said Aldo. "The fucker can't

even look me in the eye. He nods his fuckin' head. I tell him what a shitload of asswipe bosses he's got here and his fuckin' light just nods up and down. You don't get satisfaction out of these fuckers."

March 22

GRADA AND I HAVE MADE A DECISION ON THIS CRISP, BRIGHT Saturday with the sky a light blue behind the hills. Grada plays bridge every Friday night. Cards bore me. Music bores her, a minor tragedy when an ex-musician is banished to the bathroom to play his guitar so that others can hear the TV. Therefore, I will go out Saturday nights to dance and she will go to her bridge game. Since I change shifts each week, I can go only one week out of three anyhow.

Tonight there is the weekly Saturnalia at the Tippleside Veterans of Foreign Wars hall. In the past year or so, they had go-go girls, but they've discontinued them.

A dry, cold night with thousands of stars and a sliver of new moon. I stopped for a beer, then went on to Tippleside through the neon wilderness of used-car lots and fast-food huts.

From the outside, the Vets resembled a bank, with its flat concrete walls and glass doors. Inside, a man in a white shirt checked membership cards. It cost me $4 to join. Most men belong to at least one club, the Legion or the Vets, because under the Pennsylvania blue laws, public bars are closed on Sundays.

I went back to the big dance hall with its high ceiling,

glass brick windows with yellow drapes and colored neons underneath. Rows of white Formica tables surrounded the dance floor.

In my early twenties, which was the early sixties, there had been the same Saturday night dances, except that in those days we carried a pint in our sport coat pockets to spice the drinks and today joints have somewhat replaced booze and sport coats have disappeared. And back then the girls wore dresses, whereas now most of them have slacks, a regrettable change. Vodka-Squirts, still the main drink, once cost thirty-five cents, now are fifty-five cents. Many of the men still wear high-school jackets, but today the collars are hidden under long hair.

The band was an untalented group of adolescents amid ten grand's worth of equipment. They played almost nothing but oldies interspersed with polkas and obetiks. The floor was jammed with girls dancing polkas with other girls.

The big crowd drifted in around 11 P.M. Tables filled. In front, rows of stag men nudged each other and perused the stock. The ladies' room was also in front, off an alcove between the dance hall and the bar. The stags formed a gauntlet and the girls passed through it on the way to freshen up, providing a chance for the girls to be seen and to stop and chat.

I met Hippie Joe and his buddy Paul. Joe is Greenridge's unofficial hippie, a short, wiry young guy with shoulder-length brown hair and the graveled voice of the heavy dope smoker. Paul, whom I remembered as the ten-year-old cousin of a friend I had run with in high school, is now about six feet and 210 pounds. He worked at 18-D for six years and currently runs a Lee Norse at #32, high coal with a lot of six-foot roof. He said it was a different world. Hippie Joe is a buggy runner at Harken, a mine in an ultra-redneck area. He said he took a lot of grief from his crew, but he considered it his mission to "educate those dumb fucks."

They suggested that we step out to their car and smoke some dope. Just then the music stopped and the lights came on. It was time for the weekly drawing. Bad skin, the dirty dance floor, and dingy walls were exposed under the merciless glare of the neons. What had been a group of young people chatting amiably at a dance was suddenly illumined as a horde of drunken louts shading their eyes. A hefty matron, the steward, rushed to the stage with a pretzel can. One of the musicians pulled out a card, the number was read, the lights went off, the clangorous tumult from the band burst from the amps again. We moved quickly out the glass doors.

The grass made me goofy and I lost track of time until the dance was over. On the way home I stopped for a beer at Barley's in Cokeville. Inside, seated at the bar, were three women classmates from high school.

Dolly, a former cheerleader, was sipping a Scotch and water and talking with Ann and Barbara. Dolly has remained small and taut, and her pixie face, encircled by a brown page-boy, is unlined. She is divorced and works at the state school for retarded children in Arcola. She said her life now revolves around her fifteen-year-old daughter.

Ann has changed completely. From a dark, brooding adolescent, she has grown into a mature, attractive, outgoing woman. Her husband was electrocuted in an industrial accident. While raising two small children, she is using her widow's social security to attend college. She said she wants to become a psychoanalyst. About her husband, she said, "We wouldn't have made it anyhow, but I'm not happy he's dead."

Barbara, short, red hair and a sharp face with dark eyes

that sparkle when she smiles, became an unwed mother before it was fashionable, at least in this area. She is raising a six-year-old daughter with no help from anyone, barely getting by as a waitress.

We talked about the rest of the class of '58, and it seemed that most of us had screwed up one way or another. At least, those that they mentioned were unhappy according to what they had heard. We talked until about 3 A.M. when the bar closed.

PART 6

April 5

We are back after a two-day strike. We never know what goes on in the super's office, but the talk in the washhouse claimed that Adams, one of our committeemen, intended to close down an area and management wouldn't allow him to inspect it. Nova contended that they needed twenty-four hours' notice before the union could make a contract-specified inspection. So, everybody was out for two days, about $100 off the next check. It seems odd to me that we lose money because the company won't comply with the safety regulations, but that is the way it always is. Adams, a slim, solid man, crew cut, prematurely gray, aspires to rise in the union; he is thirty-five, has been in 18-D for years—some of the time as a boss—and knows the mine front and backwards.

When we went back today, Vince, the grizzled old boss with arthritis, gave me two men who had just started and told me to move two pumps. On the way to the mantrip, one of the men apparently got lost. As I was climbing into the mantrip with the other guy, a boss came by and took him away.

Undaunted, Vince took me down to A-15 and asked if I could run a motor, an electric car used to haul rail cars.

I had never operated one, but, figuring I had to learn some time, I said sure. It looked simple. We went to the end of the tracks and he showed me one of the pumps, a big 350-pound job that lay about fifteen yards off to the side. "Put that on a flatcar and take it up to A-9," he said. I told him it looked like a three-man job. He turned his light in my face and bellowed, "Ya have a come-along in the fuckin' car," then he left in his jeep.

I could barely lift one end of the pump. The come-along should have had a man at each end. But, after I got started, everything went as though God were pitching in. The pump slid out as the come-along was wound and traveled right up the steel plate which I used for a ramp up to the flatcar. Then Vince reappeared in his jeep. His expression said that he found it hard to believe that I had actually loaded the pump that quickly, but he climbed on the flatcar behind the motor and said, "Okay, now let's go get the other one."

A full car of rock dust, twenty thousand pounds, was hooked to the other end of the flatcar. I had to tow the whole business, and with Vince sitting there, I didn't want to blow it. He obviously had more faith in me than I did. I pushed the power on, and away we went, slowly gathering speed up the grade in front.

Everything was fine until we started down the other side. I hadn't thought that the load would push us quite so fast, and for a moment I couldn't remember which way to push the throttle arm to reduce the power. While I was puzzling that out, I should have been putting on the brake. We moved faster, finally flying along, bouncing on the tracks, completely out of control. Vince sat stolidly gripping the pump. Maybe he simply thought I was a fast driver. If we had gone off the tracks, we would have knocked out a slew of props and rails, probably causing some of the roof to come down, so it was no joke. When I managed to turn the brake wheel tight and the throttle off, we finally slowed and stopped. I sat there slightly shaken, and Vince growled, "You shoulda stopped three holes back, goddamit."

He disappeared into another shaft and I went back three holes. The second pump loaded as easily as the first. It was a fluke that all went so well. Since I had escaped the odds, I decided to take an extra half hour for lunch. Normal lunchtime was a half hour, the only break in the day. To escape the wind, I walked to a switch off the main shaft and crawled behind some cribs. The roof was cracked and broken, but rails were set every two feet and cribs supported much of the weight. This was the end of a mined-out shaft which was sealed and covered with discarded pieces of plank, buckled rails, scraps of corrugated metal, and anything else that was thrown off the main shaft. Long, wispy black strands of fungus hung from the rails and white, furry moss glistened on the discarded wood. The rock dust had long since turned soggy gray. Rust patches showed along the sides of the shaft where the sulphur had reacted with the coal and air.

I opened my bucket, took a drink of icy water, and settled in for lunch: three thick chicken-salad sandwiches with lettuce on homemade wheat bread. A big, red, juicy McIntosh apple. A bag of chocolate chip cookies. Two small nectarines. A piece of plank provided a dry seat and my helmet just brushed the roof.

As I was biting into the first sandwich, paper rustled about ten yards ahead beside the brattice. When I turned my light there, a fat, furry rat looked up and casually returned to his lunch of orange peels. I tossed a rock at him and he halfheartedly scampered away, circling around to the crib beside the one I was leaning against. Barely pausing, he moved back along the same line and resumed wrestling with the orange peel. I threw another rock, aimed it better, and it made a puff of dust beside him. He jumped and ran back along his same line, stopping for an instant to look into my light, twitch, and move off again. He went into the crib beside me, about a yard away.

I am not afraid of rats, but I didn't like one sitting within striking distance. As a truce, after eating the apple I tossed the core about five yards away. He ran right after it. He made a good target, so I figured the hell with the truce and fired another rock. He scooped up the apple core on the run as though he'd been practicing, made a little jump past the rock, and circled back to the crib beside me.

I refused to move. I figured I had on heavy clothes and I took enough crap down there without being pushed around by rodents. I ate a leisurely lunch, but when I finished I'd been there less than a half hour, and I couldn't stay. I spent a few minutes trying to entice the rat out of the crib so I could club him, but he knew better.

After work, I headed again toward Cokeville and Barley's. It was around midnight, misty and cold with icy patches on the road. The route seemed to be all downhill, like driving into a gloomy swamp. Street lights glimmering on hills made greenish halos through the windshield. Even with the window halfway down, the glass fogged over. The radio crackled, signaling a coming storm.

I drove into Cokeville past stinking boney piles and more tipple lights. Mansard-roofed, red- and yellow-shingled company houses lined the glistening street to Barley's. Inside, Dolly, Ann, and Barbara were sitting at the bar. Barley was not around and some kid was tending bar. He and another kid were shooting pool and teasing the local halfwit, a young fellow with buck teeth and horn-rimmed glasses.

Puck, meanwhile, was harassing the girls. Puck is certified insane, papers and all, a heavy drinker who went over the edge after his son was killed. Dressed in a white shirt and tie, he is a neat, grandfatherly man in his fifties who stays permanently plastered and whose pleasure at the moment was in molesting the three women. They jabbed him with elbows, yelled at him, moved away from him, but his ardor was undampened.

The draft beer was sour, so I took a bottle of Iron City and sat next to Dolly. The Rolling Stones blared from concealed speakers in the knotty-pine-paneled back of the bar.

"What do you girls do for sex?" I asked her.

"We lie down, usually," she said, laughing. Puck, eyes at half mast, leering, reached out to fondle her shoulder. She slapped his hand away without looking at him. She hit hard. Puck jumped back a few feet and sucked at his hand.

"No, I'm serious. What do three single women in their thirties do to get laid around here?"

The halfwit suddenly grabbed a pool cue off the table and threatened to brain the bartender. The bartender and his friend had taken the halfwit's hat and wouldn't return it. George, a young giant drinking calmly by himself, took the cue, returned the hat, and settled everyone down.

"Well, everyone our age is married," said Ann. Her

hair was tied up in a colorful scarf and she wore a black peasant blouse and jeans. "Forget the ones we went to school with. We have a choice of the younger men or the older. Natch, we take the younger ones. When we can get them."

"It's sad, really sad," Barbara said. "There's nothing. I'm the only unwed mother in town. Who's gonna marry you around here with instant family? We drink together, we smoke dope together, I mean, none of us can go away. There's no money."

Puck had gone to the rear, behind the pool table, to do a headstand with his feet against the door between the pinball machine and the dart board. Everyone stopped talking to watch the act, giggling and laughing. Puck bent over and dusted off a patch of floor with a handkerchief, smoothed the handkerchief over the patch, and put his head down on it. His feet rose slowly into the air. He bent his legs and the bottoms of his shoes clapped against the door. All of his money fell out. Change rolled around the floor, bills flopped into a heap. While the kids gathered Puck's change, Cod, a wireman at 18-D, sneaked out through another door. Puck held his position, reveling in the applause. Then Cod burst through the door supporting Puck's feet. Puck sprawled across the floor. Everyone laughed, the girls shrieked. George helped Puck to his feet, dusted him off, and handed him his beer. Puck took his pile of money off the pinball machine and stuffed it back into his pockets, grinning broadly, proud to have amused the group.

"This is the nightly entertainment," said Dolly. "He has matinees on weekends."

Several men appeared from the poker game in the back room to see what the noise was about. Among them was Zurko, draft in one hand, a paperback in the other. He went behind the bar and changed the tape from the Rolling Stones to Mantovani. When he came back, he and Barbara talked. I couldn't hear what they were saying. From the look in her eyes, it appeared that they knew each other well. Then he showed me the book he was reading by Carlos Castaneda.

"I finally figured it out," he said. "It bothered me. Now I figured out why. The sonofabitch wants to leave the human race. His whole philosophy is about leaving the human race."

It sounded like a pretty good idea, I said.

"That's just it," he said. "You don't live like that. All the honor is living it out, rotten as it is, or having the guts to kill yourself."

One of the men came out to get him and he returned to the game. After he left, Barbara said, "He's really nice, but he's so weird."

I left about 2 A.M. It was still misty and cold. At the trailer, the baby was up and crying. Grada was walking him around the tiny living room. She had large, dark circles under her eyes.

"What was it tonight, another dance?" she said, shuffling in a circle.

"Just a few beers. What's the matter, you object to my going out dancing? Why didn't you say so before?"

"I don't really care," she said. "It just seems to me that you're a little old to be playing teen-ager."

"That's right. I'm too old and much too mature to be trying to pick up girls. So I go to dance, and if you don't buy it, if you think I'm rotten enough to cheat on you, that's your problem."

"Do you cheat on me?"

"No. I can tell you that in complete honesty because it hasn't happened. If I had cheated on you, though, I'd be dishonorable enough to lie about it."

She sat down on the couch and the baby started snuffling again.

"The only reason I believe you," she said, "is that when you go out you get drunk, and when you get drunk, there isn't a girl anywhere that wouldn't spit in your face. I remember in college when your idea of a good joke was lying about your age and breaking wind in an elevator."

"Well, evidently it was enough to sweep you off your feet," I answered.

I went to bed and left her with the baby.

April 8

VERY COLD THIS MORNING. THE FROSTY GRASS CRUNCHED underfoot when I walked to the car. I was sent with Hagley to set timbers in the main shaft. The carloads of props being sent in from outside lately have been seven-foot, 150-pound logs. Hagley was pale and he breathed hard after we threw each prop out of the car. There must be a cold front moving in; the wind through the shaft was bone-chilling. The props had to be carried for about fifty feet. We put them across the tracks and rolled them for a distance, then carried them the rest of the way. To replace a rotten prop, we placed a hydraulic pump jack under the overhead rail, with the jack bottom on a couple of crib blocks. Then we jacked the rail tight against the roof and sledged out the rotted prop. Some props were tight and had to be sawed out. Hard bottom, about two feet down, was reached with a pick. We scooped out the dirt, measured the rail height, cut a new prop, and raised it under the rail. Then we knocked it in tight with wedges and moved to the next one. All the while, we shivered. Our breaths rose in white plumes and blew down the shaft. Hagley insisted on splitting the work evenly, knocking out every other prop, but it was hard for him. Every hour or so we moved out of the main shaft to warm up.

During one break, while we were sitting on oil cans off in a side shaft, I told Hagley about my lunchtime battle with the rat a few days ago. He said that in his last mine the rats attacked during lunchtime, leaping at the men's hands for sandwiches. The men would drop their sandwiches and the rats would take off with them. Finally, the miners threatened to strike unless the company fed the rats. As a result, the company brought in oats every day and scattered them through the crosscuts. Hagley said that the rats were starving because the mine was dry and the rats bred too fast.

Cooper appeared while we were in a side track away from the wind. "Whataya doin', Hagley?" he said in his snide, insinuating voice. Hagley immediately became enraged and started to walk out. Cooper saw that it might lead to trouble and backed off, talking nicely to him. Later, Hagley bitterly repeated the story George had told: that after his last sickness, when George took him home, Kurtz had docked them a half hour's pay off their sign-out time.

On our way back in the mantrip, a chunk of ice crashed into the car from the roof. Everybody jumped, thinking the roof might be caving. I was halfway out of the car. Much relieved laughter.

April 9

Up at 5 a.m., scraping the frost off the windshield. Driving to work at 6 a.m., headlights are on low beam and the fields are lit in amber and pink.

After I suited up, Crisco came over to me and said, "Wait for me at the bottom of the cage. Cooper has a job for you."

The last job he gave me was with Lou Perski, who allegedly beat his wife and threw her naked out of the house. After we went down in the cage, the job turned out to be picking muck out of the tracks in the main shaft, high and dry. Nothing to it. I was suspicious.

I had filled a quarter carload when Pumper stopped on his way to check the pumps. Pumper barreled his jeep flat out through the shafts. I could hear the iron wheels clacking and skirling and see the light bouncing from a long way down the track. He squealed to a stop and shined his light in my face, an annoying habit. I beamed my light into his eyes, but he stared back like an owl.

"How's it goin', buddy?" he yelled in his nasal whine.

We talked for a while and I told him what Hagley had said about forcing the company to feed the rats. Pumper nodded and laughed, and then he told a story about the strike before last when he had been working alone. He scabbed during strikes as well as working Sundays and holidays. He said that during the strike before last the rats hadn't had anything to eat, and had become aggressive. He had been up in a long shaft away from everything at lunchtime. When he opened his bucket, about fifty rats appeared. He closed his bucket and went outside to eat, but the rats had made him so mad, he said, that he devised a trick for them.

Chuckling and gesturing, he acted out how he had put a shovel into a puddle and laid some food on it. Then he had attached a wire to the shovel and nipped the other end onto a high voltage cable alongside the belt. When a rat touched the shovel, said Pumper, the shock knocked him a long way down the shaft.

"They squeal, Christ, you oughtta hear the little bastards." He laughed. " 'Course, you ain't allowed to do it, but I hate the little bastards. Aldo, the brattice man, he's deathly scared of rats. When a rat's around, he jumps. Oh, he's tough all right, but he's deathly scared of rats."

I asked if he knew about Aldo's fight with the state trooper. Pumper said yes, Aldo had been caught because afterwards he had bragged about it in bars where the police had stationed informers. "Old Aldo gets pretty mean when he drinks," he said.

April 10

THE ALARM JANGLED AT 5 A.M. AND I LAY LIKE A STONE, loath to swing my feet over the side. Gusts of wind rocked the trailer and rattled the windows. The furnace coughed and droned. I lay back and pulled the covers around my neck. My feet stuck out at the bottom. The bed was an even six feet long and not wide enough for both Grada and me.

The Big Ben on the floor ticked, nagging every second. Time was short. I could dash out to the phone in the kitchen, call the mine, and jump back into warmth and darkness. But then what? No work, no pay; $47.50 down the drain. And in an hour or so when Grada and the kids awoke, what vile complaints would I have to endure? And they would be right. I closed my eyes and thought about what I could do with a day off. Read the kids a book and hold those giggling little creatures in my lap. Later, take a walk in the woods with the dog. Go to the cabin, write letters. The winter hangs on. They say this is the worst winter in years; not the white, clean sweep of snow and wind that ruddies cheeks, but gray, wet, dirty slush with a dank cold that settles into the spirit. Pennsylvania suicide weather, as John O'Hara called it, that comes in December and hangs on until March. And now it's April.

The hell with it. I jumped out of bed and ran across the cold linoleum to the kitchen phone. No answer at the mine. The dirty bastards. I dressed quickly, bolted a bowl of Cheerios and a cup of coffee, took my bucket from the refrigerator, and ran to the car.

In the washhouse, suited up, I wanted nothing more than the job I had yesterday picking frozen muck out of the tracks and loading it into a car. But Crisco said, "You go with Donchus to Dutch Run."

I walked back with the crew to Dutch Run, a new section. Water poured from the roof. Slippery, oily black water covered the tracks. Donchus put me in an air shaft to timber in strong, frigid wind. Dutch Run was low and wet. Working alone made time creep by. I couldn't stop working because my body tended to freeze up like an old motor. It was hard to resume work after a break. I laid out the props in a long line. Then, starting at one end, I measured the roof with a saw and a cap piece (a wedge used to tighten the props). I sawed the prop, put a wedge on top, and drove it tight with the ax.

The roof wasn't bad, so I was able to escape through

daydreaming. I wandered back ten or fifteen years and ran through an entire afternoon at the cabin in July with an old sweetheart, a beautiful, charming girl whom I had gone steady with for six years. I went slowly through all the things we did that afternoon. She loved to wade in the creek and catch minnows in a bucket as my kids do. Later, we had walked through the woods, sat on the porch in rockers, talked, and made love upstairs on the creaking brass beds. From what I heard, she was now married and living somewhere around Philadelphia.

Working under good roof was ideal for mind-tripping. Under bad roof a man has to stay alert. On a crew, there are too many interruptions—putting up canvas, setting rails, shoveling coal. But at least on a crew the work is steady enough to kill time. Working alone makes passing time more difficult, but with practice and good roof, a man can put in a day's work and hardly know it until after the shift when the bone-weary fatigue sweeps over him in the steaming showers. My dream in Dutch Run was not enough to insulate me all day. When I had to work without it, shivering and cursing, I took breaks in a crosscut, huddled against a brattice.

At home, I didn't wash again as most miners usually do. One doesn't notice the dirt after a while: the coal ground into the hands and the dark rings, miner's mascara, around the eyes. Dust has settled into my clothes, into the furniture, into the car. The less one notices, the better. Once home, I eat, watch a few hours of insultingly stupid television, and go to bed. Each day's time is divided into ten hours for the company, including travel time to the mine, and five for me, six days a week. Each week is a different shift. My mind and body never have time to adjust between night and day.

April 11

I TOOK TODAY OFF. MY NAME WAS UP FOR SUNDAY WORK, double time, and tomorrow is Saturday, time-and-a-half. Today was bright and clear, a time to take the dog out to the cabin, run through the woods and purge the lungs of dust.

I drove to the cabin under light, puffy clouds, warm breezes. When I turned in at the heavy iron gate sunk into the stone columns, Baron, ears quivering, jumped out of the back seat and tore down the lane to wait for me at the bridge. The lane, two muddy furrows with a ridge of moss and high grass in the center, was shadowed by the pines. Soaked branches dripped into the mud. At the end, sunlight dappled the parking area and flashed off the creek. The lush greenery gave off heavy, verdant odors. I crossed the bridge, which smelled of creosote, planks slippery and bright in the sun, and walked across the pale, overgrown lawn. The cabin, squat and solid, was dark inside and dank as a tomb. After opening it to air, I dragged one of the old rocking chairs out to the porch and took the reserve bottle of Jack Daniels from its secret cache. I settled back and watched the tops of the high pines swaying in the wind and listened to the dog, yelping happily, crash through the thick underbrush. I put a pitcher of icy water from the spring on a card table beside me and prepared to enjoy a day off.

This was one of the first mild days, a promise that the hard winter would die and warm weather would come, a whiff at a time, finally thawing the cold lingering deep underground. As the day warmed to its peak, I walked along the stream, running high and clear, glittering over the rocks. By the end of the afternoon I was very drunk.

April 14

Hoot owl. I went with Hank to set seven-foot timbers in the main shaft, props with a thirty-inch girth that aggravate the bursitis in my left elbow.

The first night of hoot owl is always the same; we tried to do everything the first few hours before fatigue seeped through our bodies. Later, we took turns copping twenty-minute naps while the other man watched for the yellow gleam of a buglight, the flame safety lamp of the bosses.

Hank said he could sleep all day at home. I can never get more than four hours, maybe five, and by the end of the week I am half dead. A lot of men like hoot owl because there is less pressure for production. The people who are nervous about production work steady day shift.

We set eleven props. When the night was finally over and we were coming up in the cage, the lampman, who was running the elevator, looked down at Hank beside me and said, "You think yer gonna like the mines, buddy?" Hank just shrugged and smiled. He is eighteen and his eyes already have crow's feet, deep lines of miner's mascara. His hair was filthy, matted and streaked with black grime and white rock dust. He looked like all of us. "You gonna make a career outta this?" the lampman insisted. Hank shrugged and smiled.

In the washhouse, I showered and dressed. I was lacing my boots when Danny, the super's son, came over and handed me a small packet with a shy grin. "Here's two hits of orange mescaline," he said. "You were asking about dope before, and there's lots of it around right now."

The road was clear on the way home in the early light.

Pale blue sky, no clouds. Driving along the high ridge, I looked down at Cokeville. Rows of company houses were nestled among piles of slag and black mountains of boney. In the early sunlight, the long, brown stripes where coal had been stripped lay against the snow like the skeleton of a fish.

April 15

I WAS ASSIGNED WITH HANK AGAIN TO SET TIMBER, THIS time in the B-16 airshaft in the path of a cold, steady wind. We took the small covered mantrip back, the only mantrip in the mine with a canopy, and picked up our tools—two saws, two axes, two sledges. We carried them to the kersey battery-charging station to kill a half-hour talking with Bill Sweet and Joe Morgan, the supply men. Hank said his cousin, Danny, had gone down to see Pam at her trailer before tonight's shift. Her ex-husband, said Hank, "some big hippie with hair down to his ass," had seen him and given chase. He and Dan had argued over their C.B. radios.

"Somebody's gonna get shot over that bitch before it's over," said Sweet. "Mark my words."

Hank had been setting props with the two women last week. It was fun for them to have an eighteen-year-old there, he said. They kidded him about his lack of practice in eating pussy. Helen sat on the end of the three-wheel cart and orated on how to go down on a girl. Hank said Pam told him that she could never suck a cock.

"She said it'd make her throw up, except she'd suck Richard's cock any time. She said she'd give him the best blow job he ever had," said Hank.

Joe Morgan said that she'd told him that she had bought a C.B. radio for her car because she heard that one night Richard was so drunk that he couldn't drive. He had called for help on his C.B. "She said if he ever calls again, she'll be the first one there," said Morgan.

April 16

I WORKED THE TAILGATE OF THE LONG WALL WITH PETE, a wrinkled, weathered, short man with over thirty years underground. We were to dynamite the end off the wall so that the shearer could move in for a bite. At the beginning of the shift, after the ride up the belt, we dragged seven fifty-pound rams, awkward, bullet-shaped tubes, from the belt to our work area, about 150 feet. It was good roof and dry, with fairly strong wind.

When the shearer, the four-foot cleated wheel, arrived at our end, whirling and grinding the coal off the seam, Pete said, "You ever see anybody get caught by that thing you won't forget it."

We stayed about twenty feet from it. As the shearer ground off the end of the seam, someone back on the chock line rammed everything forward by pulling levers on the jacks, and the wheel took another bite and moved off in the opposite direction. Black dust cut the light from our lamps and made it hard to breathe until the shearer passed. Part of our job was taking down the dowdy jacks and dragging them back after each pass to set them up again in the rear for roof support. The props were sledged out and thrown aside.

I had worked with Pete once during my first month in the mines. We had been at the long wall and it was bitter cold. The roof was bad and five or six inches of water covered the bottom. As the shearer advanced, I had had to break down cribs which were used because the roof was too bad to be supported by props and dowdy jacks. I knocked out the crib supports, dug out the bottom two blocks buried under the icy water and muck, and dragged them to the rear to rebuild them. Pete had not had knee pads. Crawling over wet coal in that cold must have hurt, but he never complained. At one point, the roof collapsed over the tailgate and we had to dynamite a rail out to move the tailgate forward. The shot went off with a terrific *whang!* and shrapnel from the rail whizzed down the shaft. Pete liked that. "It's just like a fuckin' war, ain't it?" he had said with a toothless grin.

Today during lunch, he was excited about the summer of '36, when he caught some striped bass and trout from one of his favorite streams. He had transplanted them to another place, he said, where they had grown to amazing lengths. He wasn't in a coal mine at all during our lunch; he was young and happy, sitting beside his favorite trout stream. Afterwards, he put his dream aside along with his empty lunch bucket and we went back to work.

April 17

I WAS SENT WITH A CREW TO HELP MOVE THE BELT FROM one section to another. When an area is mined out, all the belts and equipment are moved somewhere else.

The phones were out. At the tail, we didn't know what the men at the head end were doing. The first step was taking out the rollers, leaving every fifth bottom roller to support the belt. The rollers were thrown on the belt.

Then the belt was clamped tight and turned on. The tail was dragged forward as the belt took up the slack. The tail, a large hunk of flanged steel, flew up the shaft at 250 feet a minute, the belt slow speed. If something happened to deflect the tail off course, the belt just kept pulling it, through rows of props, bringing down rails or roof, until it could be stopped. Endive, the dwarf, was used as a runner between us and the men at the other end of the belt. It was hard, wet work that took half the night.

After we had the structure out and the tail moved, we had to move the belt itself. The easiest way was to haul the end of the belt out into the main shaft, then attach it to a motor car that would pull the belt up the shaft. After we had done this, the boss said, "Take it out fast so it lays smooth."

The man on the motor nodded and took off. The belt caught on an overhead rail and yanked it down onto a high-power cable. The cable broke and fire flashed through the shaft.

The yell of "fire" in a coal mine is terrifying. There was little danger of an explosion because 18-D has no methane to speak of, but the cable insulation gave off noxious fumes, and the rubber belt was burning. Someone arrived with a fire extinguisher, but it didn't work. Another extinguisher was found and the fire was put out just as the first jeep full of men from outside sped down the tracks. Everyone was excitedly talking at once, explaining to each other how the belt should have been taken out, where the other fire extinguishers were, and how to take belt out in the future. Finally the cable power was shut off and we went back to work.

Later, I was riding up the tracks with Mort, one of the face bosses. He is fifty-one, four years away from retirement. He talked about going on vacation next week up to his hunting camp. "Yeah," he said, "I like going up to camp. Nobody gets to you. This place here, you make money, but fuck, you don't get time to do nothin'. You retire and huff and puff, what's the fuckin' money do for you then?"

April 18

WE ARE ON STRIKE BECAUSE THE ELEVATOR IS BROKEN. The company wants us to walk down the slope. The union says no, that there must be more than one exit in case of disaster.

April 19

SNOW TODAY. WE ARE STILL OUT. AROUND THE TRAILER camp, strike behavior is moving in; nightly screaming arguments, loud music, sounds of smashing glass and screeching tires. Daytime, cars sit on blocks with men underneath, usually with a six-pack killing time by adjusting the brakes.

Billy Sweet and his wife stopped by today. He came in for a beer, left her in the car. We talked about the strike, and he said he didn't have the rent money for this month.

"People think miners make a lot of money, but we're out so much, shit, you can't get ahead," he said.

In the afternoon, Grada and I packed the kids into the car and went shopping in Pillartown. In a new mall lined with boutiques, I saw a denim outfit, pants and jacket, that looked secondhand. A tag on it read, "This garment is sewn and scientifically laundered to give the look of being old and worn. Flaws and imperfections are part of the total desired look." The price was $60.

April 23

We have been out for a week. At the union meeting, the vote was to go back to work yesterday, but the mine didn't call the men on that shift, so no one went back. Today we all showed up, but Adams, the committeeman, said the union's position was that with only one way out the mine was unsafe. We sat around the washhouse and read *Playboys* for an hour, then left.

At the River Gap Hotel, we stopped for a few beers. The bar was jammed. No one wanted to go home. Sweet and I decided to make a day of it.

At around 3 A.M. we ran into Hank and a friend of his. They were drifting around Tippleside looking for girls. All of us had started drinking in mid-afternoon. We sat outside a bar around Hank's new GMC truck, all gussied up inside with big eight-track speakers, C.B. radio, and other gadgets. We had a few beers, then Billy and I went to Cokeville to drink.

April 24

There was some excitement in Cokeville after Billy and I left last night. A man went berserk, shot two cops who had come to evict him from his rented house, and then killed himself. His house burned. The police attributed the fire to tear gas shots fired inside to bring the man out.

Larry, our neighbor in the trailer above ours, said this afternoon that there was a bomb threat at his mine yesterday. He works Nova #37. Larry is trying to get mortgage approval from the Greenridge bank for a double-wide modular house. It's his old lady's idea, he says. He likes the trailer park. On nice days, he sits with a beer on the stoop of the front door or out on a lawn chair and watches everything. He said that where his old lady wants the house, he doesn't know anybody except for one buddy, and the buddy is on a different shift.

April 28

THINGS ARE SERIOUS. WE HAVE BEEN OUT FOR TEN DAYS, and the rent, car payments, and other end-of-month bills are due.

I went to the union meeting today. The long, green hall held only fifteen irritable men. Tempers were short. Adams, the committeeman, said we were out because the cage was down. Normally a broken elevator can be fixed easily, but he said ours was Japanese and the parts had to come from Chicago. Someone suggested that the fifty thousand tons of coal piled outside the tipple might have something to do with the layoff, but Adams didn't think so. "Of course, you never know with these people," he said. "They lie every chance they get."

I thought that it must be the elevator because the company wanted us to return to work. But they wanted us to walk down the slope, and Adams said no dice, it was against the law; there must be more than one exit from the mine in case of disaster. He said that the company claimed that there were actually three exits: out the main belt shaft to the tipple; the slope; and a shaft that led down to the B seam in another mine. No one I spoke to had the vaguest idea of how to find the B seam shaft. Adams said that a boss had been hurt last week in Main A and it took them two hours to get him out. "That's a lot of time to die in," he said. If we were called by the mine, we were to go to work, said Adams. We would suit up, but not go in.

The mine called when I got home. The lampman said to report tomorrow, and if I did not go down the slope, there would be no shape pay—the few hours pay for showing up when there is no work.

April 30

I WENT TO WORK. WE SUITED UP, THEN EVERYBODY SAT around the washhouse. Superintendent Kurtz called a meeting. Standing in front of his office in his thermal underwear, he said, "Men, we have three escapeways out of this mine. We have the slope, and out along the belts, and down a temporary shaft to the B seam. Now, the buses are outside, the mantrip leaves in five minutes. That's why we called you men. We expect you to work."

We went back inside the washhouse. A few baskets swayed from the ceiling. Men sat on the benches reading *Playboys* and talking about pussy. There was no discussion of the strike. That issue would be decided by Adams, and the rest of us simply awaited a verdict. Then Adams got on a bench and everyone crowded around.

"I want you men to know that we have the backing of the national, and our lawyers expect to win. The company says it's safe down there, we say it's not. It's as simple as that," he said.

He stepped down and the men returned to sitting on the benches. I went outside to the parking lot where men hunkered and leaned against cars and trucks. The bus in the middle of the parking lot roared its motor, doors open. No one looked at it.

Pumper told me a story about when he worked at #50 in Cokeville. "There was nine million gallons of water pumped out of there every twenty-four hours. Can you imagine that? The men were from seven to twelve miles up in the shafts in that place. We had to walk out once. I was the last man out. I walked all seven miles, buddy."

I didn't hear anyone say it was time to go, but men in the washhouse started changing clothes. I went inside and changed and asked if anyone knew when we would go back to work. I was told that we would find that out at the union meeting. No one in the washhouse debated about going back, probably because to espouse going back would be highly unpopular whether right or wrong.

I went home, feeling morose. The story on the back page of the daily paper read: FIRST QUARTER PROFITS SOAR AT BETHLEHEM, U.S. STEEL. It said Bethlehem's profits were nearly double over the first three months of last year. U.S. Steel reported earnings had more than doubled from the first quarter of 1974, despite a 25% drop in shipments.

Apart from that, U.S. Steel Chairman Lewis W. Foy said that the recession was seriously affecting many segments of the steel market. It made me proud that Mr. Foy and his group, along with the owners of 18-D, could make such a killing in the midst of all these wildcat strikes.

Meanwhile, it is still cold, gray, and wet at the end of April, and we have been out for two weeks, $500 gone from our pockets. The only recompense is unemployment. With a family of five, I am eligible to draw $119 a week. Nova, however, regularly disputes every claim and we won't know whether we are eligible until after a court decision.

It is impossible to stay in the house when we are on strike. I left in the late afternoon and went down to the union hall to see if anyone knew when we were going back. In the warmer weather, the slag heaps along the Cokeville road had turned bright saffron yellow and orange, mixed with coal black.

In town, I turned right and passed under the stone trestle to the club. There were only a few men in plaid hunting jackets sitting around inside. Zurko was at the bar reading an old Baltimore Catechism. Tom, the bartender, said nobody knew anything yet, but there would be a meeting soon.

Tom and I used to have a lot of fun together when I was a teen-ager. I played an old upright piano in the back of the club, and Tom played the ukulele. Six feet three and 240 pounds, he was too big for the uke, but he cradled it and strummed away like some giant Arthur Godfrey. Each finger covered a couple of strings at a time. After a few pitchers of beer, cradling the uke against his big chest, he sometimes reached for a difficult chord and snapped the neck off the uke.

The piano was gone now. Tom wheezed with black lung and tried to make ends meet by tending bar. We hardly mentioned the old days. Zurko, meanwhile, looks pale and underweight. He showed me a picture in the catechism of a signpost with arrows pointing toward a flaming hell. The arrows read: Love of Riches, Love of Pleasure, Love of Fame, Love of Power. Pointing to Heaven was one sign: Love of God.

"See," he said. "They make sure from when you're six that it's in your head, see, if you ever want to amount to anything you got to feel guilty about it. Give me the child and you can have the man. Goddam Nazi philosophy."

I asked him why he didn't desert his wife and kids, sell the house, and move away, whatever he wanted to do. He replied that he could never get enough for his house and that he was too old to walk out on his family and start life over, especially with no skills and no friends anywhere else. Besides, he added as an afterthought, it wouldn't be an honorable thing to do.

"It went by too fast," he said. "They set me up and faked me out before I knew where my goddam life went."

It surprised me that he thought his life was over at thirty-eight, but he had evidently thought about it more than I had. He was setting up beers for me, but I left before it turned into a night's drunk.

May 5

THE CAGE HAS BEEN FIXED, AT LEAST TEMPORARILY. I went back to work on hoot owl. The weather has broken and robins have arrived. Last night I was back at the track in the high, picking out the muck. Pumper stopped again on his rounds. He said that in thirty-eight years underground he had been hurt only once and that was a cut hand. So, not everybody gets injured badly. He said that we should not have been out because the cage was down, that Adams faked the issue because he hated the district. Back in '69 when the United Mine Workers' reform movement—the Miners for Democracy—was strong, Adams had run for an office and the others had lined up against him, Pumper said. Adams had lost and now he was bitter.

May 6

LARRY, UP THE HILL, HAD HIS $20,000 MORTGAGE APPROVED for a double-wide modular house. His mine, #37, was out three unconsecutive days last week and three days the week before. "How can you save anything when you work part time?" he said.

Night shift. By day, the trailer's thin shell echoes with squalling children. The tin walls keep out nothing. I can't sleep. Light shows through the blankets over the windows. Kids run around the trailer playing cowboy.

Last night I was back on the track, picking out muck. It was not a bad job, but it was next to the mechanic's shanty where the boss hangs out. I couldn't even sit down often, much less grab a nap. By the end of the shift I was staggering to the car with a shovel of muck.

A half hour before quitting time, a mechanic came by and yelled for me to jump into his jeep. "Bring your shovel!" We flew down to the head end of one of the main belts. The belt had become packed with coal and stopped. I shoveled the coal out with him, then he took off and told me to stay awhile and keep an eye on things. I promptly fell asleep.

I awoke suddenly, terrified. It is nightmarish to awaken

underground: the clatter of the belt, water dripping from the roof, everything black, rusted, foul-smelling, rats darting around. The shaft looked like a dungeon, cold and full of filthy pools of water stinking from sulphur. I lurched through the shaft under the low roof streaming water and crawled through the low sections, my adrenals churning. After ten minutes, I was totally lost. I had taken off up the wrong shaft. It took another twenty minutes to find my way out. I walked to the cage and joined the last few men going out.

At home, the kids were up and Grada was cooking breakfast. I slumped into a chair and looked at the TV. A Mad. Ave. suburbanite was dancing around his wife telling me it wasn't enough to *be* clean, I had to *smell* clean. Somebody out there is making $50,000 a year to insult me. I turned it off and the kids started screaming, "Captain Kangaroo is coming back." I watched a little more, and found that I liked Mr. Greenjeans. His life didn't look like any picnic either.

PART 7

May 8

MEMO FROM THE TRAILER PARK MANAGEMENT TODAY: Watering lawns and washing cars is prohibited. Immediate eviction if caught. Their well does not produce enough water and the management is too cheap to sink another one. They seem to think it's just a matter of issuing an edict to convince the tenants that their water is a gift from the management dependent on good behavior and not included in the rent. Oddly enough, most people here seem to accept that.

May 10

WARM, SUNNY, BLUE SKY, SPRING WINDS, ROBINS. THE KIDS built a tent in the yard, a swatch of pegged canvas over the clothes line. On the hillside, aluminum curlers glint in the sun and lawn chairs sag with white bodies.

The blackboard at work said, "No Work Saturdays Until Further Notice." Today was Saturday. No more time-and-a-half. I went into Greenridge under the maples budding along Main Street, to the Legion. Softball teams were being organized.

In the evening, I went to Tippleside to the Vets with Hippie Joe and Zurko. It was crowded with young miners and others standing in back. On the left, a row of tables was pushed together, filled with the long wall crew and their wives. The band, three talented hippies in tank shirts and beards, played loud and well. I drank Vodka-Squirts.

While the music crashed off the walls, the floor filled and everyone convulsed in spasmodic jerks, ignoring each other. When the music stopped, they remained rooted to the floor muttering small talk until the next number. I went out to the bar, past the popcorn machine. Zurko called me over and said, "Suppose a great astronomer does not know why God made him and a little child does? Which is better off? Why?" He was reciting from the Baltimore Catechism again. He didn't look well. I asked if he was sick and he said he was dying.

May 13

Clive, basset-faced and baggy as ever, and two new men and I were sent to rock dust—to throw white pulverized limestone against the sides and roof of the shaft.

The two new men were in their first week. Johnson, a short, dark, eighteen-year-old, and Glover, a big, overweight young man with a Hitler mustache. Before we went up the shaft, we had to unload a full car of rock dust, four hundred bags at fifty pounds each. I tossed them out of the rail car to Clive. He tossed them to Johnson, who threw them to Glover, who threw them on the belt. The long wall crew was taking them off at the other end.

Emptying a rail car is awkward, hard work. The roof was too low to stand, and after a few dozen fifty-pound bags, my back began aching. I had to either kneel, which made the five-foot throw arduous, or stoop, which tugged the lower back muscles. Clive grunted every time he caught a bag. At first, he was mumbling about "sum-bitches, we got to unload for all three shifts," but after a while we were all sweating bucketfuls and there was no talk.

After the bags were sent up, we piled on the belt. The kid in front of me, Glover, put his head down and acted like he was sleeping. I shook his foot and told him to watch where we were going, you never know when you'll have to bail out.

Rock dusting was easy, but our lungs filled with dust. The water in my bucket filled with gray sludge and my mouth became cottony. We split up, Clive and Glover at one crosscut, Johnson and myself at another farther down. Johnson said that he was getting married next Saturday. I asked him why, at seventeen. He shrugged and said he had to. I asked why he didn't consider an abortion, and he said no, they were planning on marriage anyhow, it might as well be now as later. I asked if he intended to make a career of the mines. He shrugged again and said, "Kind of looks like it." He has a brother on the long wall, and Elwood, the long wall boss, Danny's brother and Hank's cousin, was married to Johnson's sister.

The shift went fast enough. An advantage of working with new people is the fresh fund of conversation for a few shifts. After that, it is all hunting, trucks, and the weather.

On the way home, I stopped at the fruit market. Danny Kurtz was there, eyes limpid and wide. He had been to an all-night party at a local dam. He asked if I wanted to buy any acid. "I don't know how good it is," he said, "the dude that bought it has ninety hits, but I took a black beauty and I don't know what I got off on."

May 14

ARGUMENTS AT HOME. I BITCH BECAUSE GRADA COMES home from the market in curlers and a scarf and goes to the laundromat in jeans and my T-shirt. Her response is that I got us into this, and she wouldn't look like that in the suburbs.

It was foggy and rainy today. I was assigned to rock dust with Hal, the twenty-year-old supply man, and Knopick, the skeleton with the nine-inch dick. They went to the work site on a personnel carrier, an electric three-wheel cart. I went up on the belt. I couldn't find the site. I proceeded about a half mile to a brattice with a sign DO NOT RIDE THIS BELT, so I got off. It turned out that Hal and Knopick were across from me in the air return shaft. A buggy had broken down and they couldn't get the personnel carrier by. Had I continued up the rest of the way, another thousand feet, I would have been waiting for them all shift.

We had to move the bags of rock dust with a three-wheel cart, twenty-two bags a load. Low roof, slippery bottom. It was hard to pull over eleven hundred pounds a load through there. Finally we hit a water hole that ran for two crosscuts. A pump was broken. Common sense said we couldn't push a cart of rock dust through a water hole that size, but the boss would not be by for four hours, and we had to do something. We decided to give it a try. Hal was in front, pulling the long handle, and Knopick and I were pushing from behind. We pushed extra hard to sustain the momentum when we reached the water, and we went into the hole until the water rose over Hal's boots. He tried to stop the cart by raising the handle and jamming it against the roof, but he didn't pull his hand out in time. His fingers jammed into the roof with the handle. He screamed, and we pulled the cart back and waited while he eased the glove off. Blood ran from his fingers, slashed by the jagged rocks in the roof. He was able to move them, nothing broken, so he didn't go out.

We pulled the cart from the water and strung out a hundred bags of dust end to end on each side of the shaft. Hal clotted his cuts with the rock dust. We slit the bags and dumped three at a time in the center of the shaft. Knopick went ahead, cutting and dumping the bags, while Hal and I moved along behind him and threw the dust against the sides and roof.

Hal's wife is a senior in high school. He has been in the mines a year and a half. When he married, his wife was a junior, and he is taking Friday off to go to her prom. Hal said his father had started in the mines when he was fifteen. They had worked a sixteen-inch seam, lying in water and picking out the coal. They had to crawl the shaft on their bellies, he said. Now his father sits home with black lung and can't go anywhere. Hal said that his wife, being young and horny, was upset with him because he was often too tired to make love. He blamed walking in the low; said it was the hardest work that miners have to do. His father once talked it over with a doctor, Hal said, and the doctor agreed that walking in the low for hours a day affected a man's sex life, though he couldn't remember the doctor's reasons. I said that I had heard other men say the same thing. My sex life was lousy, too, I told him, but I didn't know whether it was from walking in the low or seeing my old lady in curlers.

Going out, we were last in line for the cage, so I took my time. I had a long, luxurious steaming shower, snorted out hunks of mucilaginous black snot, and hung my clothes carefully so that they would dry. I hated to suit up in wet clothes.

It smelled delicious outside. Frogs croaked around the settling ponds and the woods were alive with spring life. Warm, humid patches of fog drifted over the road. I put down all the windows and drove slowly, watching groundhogs lumber along, rabbits dart down the path of the headlights, and listening to the chirps and chatters of spring in the fields.

May 15

SECOND SHIFT. I LEAVE AT 2 P.M. AND GET HOME AT MIDnight, if I come straight home. The kids rise at 7 A.M. Not just our kids, but the entire trailer camp's. Since I can't get to sleep until 1 A.M. or later, I'm always tired.

I was back rock dusting with Hal last night, along with his usual buddy, Taylor, a very thin, pale blond eighteen-year-old who described himself as a real nut who would do anything. Taylor has a motorcycle with which he performs stunts such as riding over walls on ramps. He said, "If I ever crack up, I'm going to do it right, 'cause I'm not coming back with one leg." He said a friend of his had killed himself last week in a Corvette at 110 m.p.h. He seemed to picture it as a romantic way to go. Taylor's father is the local cop in a redneck town near Hippie Joe's mine.

I made a mistake. Vince, the arthritic old section boss, said to go back up the belt to the end and dust that section. Taylor and I were to ride back and pull the bags of dust off the belt. Hal would be loading the four hundred bags on from the head end. On the way back, I noticed the shaft was dusted already and assumed that the previous night shift had done it, dusting forward from the tail end. So, Taylor and I got off the belt, walked back two rooms (crosscuts), and started pulling off the bags and laying them end to end along the belt. When Vince came up he was furious. We had unloaded them at the wrong place; night shift had only done a small portion.

"Why don't you fellas do what you're told? Why can't you just let me tell you what to do? Now the belt's gonna be forwarded and how are you going to get these bags back to the tail?" he said in a pained voice, shaking his head. We were sitting on the bags next to the belt. Hal arrived. We could see his light shining off the rubber trough of the belt several hundred feet down the shaft. When he stopped and got off, he was coughing in long, hacking spasms.

I suggested to Vince that the only way to move the bags was with the three-wheel cart farther up which Knopick and Hal and I had used last night. We knew it wouldn't work because there was not enough clearance between the props and the belts in some places, but it was the only thing to do. Vince just nodded and went away through the next crosscut. He didn't want to know that it wouldn't work.

We stopped the belt, dragged the cart over the top,

dragged a load of bags through a canvas check (a curtain of yellow canvas used to deflect the air), and loaded the cart. We dragged the cart until it wouldn't go any farther. Hal had to stop continually, racked by coughing fits that left him leaning against the props. He said he had gone to the hospital. They gave him a shot and prescriptions for $50 worth of drugs that he couldn't afford to buy. Drugs are not covered by miners' welfare.

Hal said his wife had had tuberculosis when she was nine, and so had her younger brother. They caught it from their father, who had picked it up "from a nigger in Cleveland." He had been hospitalized for a year, and just last week he was quarantined in his bedroom because the doctors thought it might be coming back. Hal had visited him and asked if he wanted anything. The old man had asked him to go for coffee.

"I go, sure, why not? And I go down and get two coffees. He takes sugar and I don't, so when I get back he takes a drink and says, 'Do you have sugar in yours?' I drink mine and tell him yeah, and we switch. Ever since then I've had this cough and it won't go away."

Taylor also had a cough. He said he had been underground since February and really liked it. He intended to make a career of it. At his physical, the X-ray technician had told him his lungs looked like he should be retiring instead of starting in the mines. He is a heavy smoker and a volunteer fireman.

We dusted a couple of hundred feet until Hal had to stop. His chest was hurting. We spent the rest of the shift stacking the bags so they could be moved up where they belonged on the next shift.

Vince stopped again toward the end of the night, returning from his runs up to the long wall. He wanted to raise some more hell with me, but he was wheezing too much and finally gave it up. He sat on the bags with us for a while, drawing diagrams in the dust to show us what to do tomorrow, and left.

Taylor was taking first-aid courses recommended by the company, but Kurtz refused to let him change shifts on the days of the courses. He said that so far it had cost him over $600 in lost wages, but he wouldn't quit. Hal said, "Shows you how much the company wants first-aid men, right?"

Hal said he was looking desperately for a way out of the mines, but unless his father-in-law could teach him carpentry, there was nothing he could do. He said he refused to raise kids in a city, and without a skill, there was no work in the area.

May 16

COLD, RAINY. I WANTED TO TAKE THE DAY OFF BECAUSE my right knee was swollen and sore. I could stay home with the thought of relaxing, reading a good book in the easy chair. Maybe tell the kids some stories, watch a movie on TV.

Grada, stirring a pot of spaghetti sauce on the stove, said, "Your mine is down to five days a week, now. We can't afford to live on four, but do what you want, I have nothing to say about it."

Another lousy argument, ended with my departure for town. Hadley ran outside after me, tears streaming down her face. She held out her arms, shouting, "Daddy, don't leave! Please! Don't go away!"

I took her back inside, heartsick, and then continued into town, hating the fighting, the destruction of children. It seemed to me that Grada's ideas were horseshit middle-class feminism picked up from women's magazines. She seemed to think she was oppressed and it was up to me to get her out of it. I thought about the summer we had spent in northern Spain, where Grada's father was raised. We had lived in her godfather's house and marveled at her relatives. All subsistence farmers, dirt poor, and the happiest people we have ever run across. Each seemed to live for the moment. They knew exactly who they were and had love and respect for each other. If one of them was successful, everyone rejoiced in it. If one was not successful, he was still loved and respected. Money worship had not yet permeated their lives.

Halfway into town, I suddenly turned around and went to work without my lunch bucket. I spent the day sliding around under thirty-six-inch roof, throwing props in cold muck while water dripped steadily.

May 20

DAY SHIFT. I WAS SENT DOWN TO A SLOP HOLE WITH George, the barrel-chested old timberman who broke me in. We put on wet suits and taped the bottoms around our boots. George was working in a foul-smelling hole with four inches of scummy water. The roof wasn't bad, but it was no more than a yard high anywhere. The water encouraged fungus. It hung everywhere in long, red slabs and black wisps almost like curtains. I brushed my face against a red piece and gagged. It was like a fresh squirrel pelt. I dragged the props, two at a time, slipping a noose of wire around them, and George set them. He had to dig down through a half foot of muck under water without being able to see the hole. The wet suits didn't help much. We were both soaked after the first hour.

At lunchtime we went outside in the main shaft to escape the stench and water. George talked about his daughter. He spoke in slow, measured lines, making an obvious effort to be fair about everything. His oldest daughter was six weeks from her college graduation and had just announced she didn't want to finish school. He said she had fallen in love with a no-good young man and wouldn't listen to her parents anymore.

"I have good morals. I believe in a decent home, and I know things aren't the same with young people today. I don't say they're entirely wrong. But here she went all this way, and she doesn't even know why she wants to quit. She doesn't know what's wrong with her, and we don't know either," he said.

George doesn't drink or smoke. He believes in hard work and fair play, that a man's job is to provide for his family, and the family's duty is to respond with love and respect. His daughter's behavior didn't fit with that idea, and since George had spent years in slop holes sacrificing his life for his family, his daughter's irrational behavior annoyed the hell out of him. His argument with Cooper, the foreman, which had put him on Cooper's shit list and kept him on it for over a year, had begun when Cooper made the unforgivable error of criticizing George's work.

The belt that ran through the main shaft went off, and Adams, our former committeeman, came down to fix it. He was elected president of our local last week by a comfortable margin. I thought he had been a capable, aggressive committeeman, and I voted for him. His opponent was an older man whom I don't know. Adams is a beltman, and being local president doesn't alter his work schedule.

He told me that by becoming local president he would make about $50 a month, compared to the $70 biweekly paid to local committeemen for union work. Committeemen are elected annually and local presidents every two years. He said that if he was successful, his next step up would be District Council, a full-time union position which would take him out of the pits. Members of the District Council and their representatives, the International Executive Board, who meet in Washington, D.C., are elected every four years.

I asked Adams how he was going to deal with the company now that he was local president. He said he was a bit more radical than the older fellow he had run against, but that he wasn't radical as long as the company obeyed the law. "When you're in the mines, you get into management or the union. Otherwise you end up another wasted miner. They use you up and throw you away. You got to make your move early. Nobody thinks they're going to stay here. I've been underground for twelve years. When I started, I thought, well, just for a while because things are tough right now. But you get stuck. Lots of guys never thought of it as a career."

When we came out at the end of the shift, George hung his wet suit outside the door of the small shaft. I decided to keep mine on and rinse it off in the showers. Going back in the mantrip, George said, "You know, I've worked in water for years. It never bothered me, but now I think I've had it. My joints ache lately, goddam really give it to me."

As we emerged from the cage and were hanging up our lamps, putting them on charge, I glanced into the office and saw Cooper leaning back in a swivel chair with his feet on the desk, looking around and stroking his mustache with that vague, far-off expression.

May 22

A WARM MORNING MIST HUNG OVER THIS HUMID SPRING day. Instead of working with George, I was assigned to my old job of picking muck out from the tracks and shoveling it into a car. It had washed down from the long wall and filled in the tracks again. About a third of the way through the shift, I carried the pick to the mechanic's shanty to sharpen it. An inspector came by with Crisco and told me to bring the pick over to the other side of the tracks. He handed the pick to Crisco and pointed to a pothole in the roof about three feet across. "I saw three men killed by one of these niggerheads not too long ago," the inspector said.

Crisco chipped out the pothole while we watched. The inspector told stories about potholes five feet across and larger. Then he came back to the story of the three men. "They worked on that niggerhead with wedges, rock bars, every damn thing, but they couldn't pull it down. Then they went to lunch, and when the crew came back to work at the face, the fuckin' thing fell out and killed three of them."

After Crisco pulled the pothole down, I took the pick and returned to the section of track I had been working on. The motorman hadn't brought a rail car for me yet, so I stopped to talk to Joe Morgan and Pam. They were unloading cinder blocks from a car and putting them on a skid so the kersey could tow the blocks up to the long wall. Helen was going to quit, Pam said, because her boyfriend, a buggy runner in another mine, had heard rumors that she was dating men at this mine. "I don't know what his problem is," Pam said, "she never says nothin' when he brings his women home."

One problem seemed to be that Helen's boyfriend's previous girlfriend had been killed in a motorcycle accident. Her father works at 18-D on another shift, and resented Helen's taking his dead daughter's place. He had started the rumors, Pam said.

Pam said she wasn't feeling well, so I unloaded the blocks with Morgan. Then Strecke, a crusty old motorman, arrived with my car. He hates Crisco and says so often.

"Hey, Pam," he said earnestly, "I'm not being smart or nothin', don't get me wrong, but why don't you take Crisco up there around the corner and throw him a little somethin'? Maybe straighten the sonofabitch out a little."

She graciously declined. Strecke was pushing a string of six cars loaded with crib blocks, props, rock dust, and planks brought in from outside. He was able to wear a single sponge jacket with a hood now instead of wrapping himself like a mummy against the driving, icy wind that whipped down the slope in winter.

Later, while I was shoveling muck into the car, he stopped again and yelled, "Hey, wouldn't ya like to throw some cock inta that little woman? Boy, I sure would like to."

May 23

THIS MORNING IN THE WAITING HALL, CLIVE WAS PACING outside Kurtz's office, mumbling, working himself up to go in and complain. I asked him what was wrong. He said the inspectors had shut down part of Main A yesterday and the mantrip couldn't go through. The men had had to walk from the closed section on down, about a half mile for some of them. On top of that, the belts had gone down where Clive was working, and he had had to crawl that half mile.

"Sumbitchin' cocksuckers, I'm s'posed to be a fuckin' timberman. I had to crawl outta there, both them belts went down," he said.

His eyes were puffed into slits. He said he couldn't sleep because the crawling had inflamed his arthritic knees. He shuffled back and forth rehearsing his speech for Kurtz, spitting snuff into the garbage can, hitching up his pants, trying to cover his nervousness with anger.

He had told me once when we were setting props that the best time of his life had been in France during the Second World War. His face had relaxed while he talked, and smiles had replaced the squinting scowl. He had laughed, recalling those few years when he was young and important in a foreign country, drinking wine with young French girls in the hills outside of Paris. From there it was into the pits; thirty years gone in a flash.

I was sent with a seven-man crew, including the two women, to drag belt structure from the abandoned long wall. We rode on a skid, a slab of metal attached to the kersey by a heavy chain. Because of low roof and two large water holes, we used props and cinder blocks to kneel on. I also used my lunch bucket. In some places we had to crouch almost flat. I turned my head once and water from the roof ran into my ear; it was as if an icicle had stabbed into my brain.

About four inches of water clear as a trout stream ran through the work area. The pumpers were routing the water to a dam. We couldn't sit or kneel. Hunkering made my knees begin to ache. The girls started arguing; "Why did you tell everybody I was dating so-and-so," completely disgusting the rest of the crew. The boss, shaking his head, sent them ahead by themselves to loosen bolts on the structure.

I pulled structure from under the belt with a new kid, John, a long-haired eighteen-year-old who complained that the women should take a turn pulling the structure out. Each piece was imbedded deep into the mud, and the belt on it was heavy. We planted our feet, squatted with the structure between our legs, concentrated for a moment, let out a loud grunt, and pulled with everything we had. The man on the opposite side did the same, except that he was pushing. The place sounded like a karate dojo. After two or three attempts, the piece might come loose, or it might take ten or twelve tries. After each piece, we rested for a few minutes, gasping, sweating, hands trembling, and then we went on to the next one. An hour seemed like a day's work.

"The women make the same fuckin' rate we do," John kept grunting. "They should do the work or get the fuck out."

We ate lunch in a crosscut out of the cold wind. A beautiful translucent curtain of amber and white fungus hung from a plank there. John and Helen, the blue-eyed Hell's Angels' mama, shined their lights through the fungus and talked about tripping underground. Either they had slang terms I had never heard of or they used drugs unfamiliar to me, but I couldn't understand a word of it. So much for being hip.

After dinner I went into Greenridge to the Legion and met Hippie Joe and Zurko. It was Friday night and the place was crowded. Joe and Zurko and two other miners,

Chunky and Bill, were headed for a dance at the Springer Hotel in Fairview, so I went along.

First we stopped for a couple of six-packs. Chunky turned on his eight-track, Hippie Joe pulled out his pipe and baggie of dope, and we turned down the Old Plank Road into the country. Hippie Joe was telling a story about the past week when he had gone to work so screwed up that he stood beside the cage for ten minutes knocking himself on the head with his lamp before he realized that he hadn't put on his helmet yet. He was in a good position for working while fucked up, he said. He had a short, straight run with his shuttle car from the Lee Norse to the belt.

We turned onto another back road to Fairview which wound under a tunnel of giant maples. Chunky slowed to a crawl. Warm breezes wafted through the open windows while stars and a bright half-moon lit the fields and open sections of the woods.

When we arrived, we parked in back of the Springer, a piss-elegant old hotel. We could hear the music from fifty yards away. Inside, we paid a dollar cover at the door and moved into a very crowded room with large, heavy mirrors, orange lights, and crashing music. The bar was a small circle with a blue canopy of tiny stars overhead. Beyond, a small fence separated a postage stamp dance floor on one side and rows of tables on the other. The dance floor was jammed. College kids from a neighboring town mixed with the miners. The only difference I could see between the kids and the miners was that the miners were better built, more screwed up, and more aggressive with the women.

May 27

Night shift of Memorial Day. There was only a skeleton crew, maybe a third of the regular shift. Hank and Danny Kurtz were working, both back from separate trips to Virginia Beach. Hank's eyes were baggy and he squinted against the glare of the washhouse neons. He said he had slept eight hours of the last thirty-six. His cousin Dan was slumped against the doors in the hall. He had gone to a van festival in his truck somewhere around Virginia Beach. Dan never took maps or directions; he just jumped into his van and tore off, rapping with the truckers on his C.B. to find shortcuts and avoid speed traps. He said there had been a "bazoom" contest at the festival.

"All these chicks got up on the stage and showed their tits and the guy went behind each one with a trophy while the crowd clapped," he explained. "I was one of the dudes at the front of the stage who told the other people to stay back."

I went down to timber a cave-in with Hank. It was a major one, about a hundred feet long and fifty feet high, which had caused the section to be closed a few days ago. The mantrip ran directly through it. We climbed up the cribs to the first floor of planks across the bottom. Rails had been laid across about every eight feet up. Hank immediately laid down on a plank and fell asleep. We stayed back against the cribs because small chunks of rock were falling from somewhere over us and smashing below on the tracks.

We weren't there long before Hudak came down the tracks in his jeep and ordered us out. "T-t-t-timber down below," he said, so we climbed down. We had to unload another car of huge seven-foot props. They must have weighed well over a hundred pounds each. We unloaded a half dozen and Hank fell asleep again. I piled some bags of rock dust into a heap and tried to sleep, but it was too cold and wet. I finally dozed off and awoke at 5:30 A.M. so stiff I could hardly move. Hank was almost paralyzed from lying in the cold dampness. If we kept this up long enough, he said, we'd probably get arthritis of the spine.

May 28

I SHOULD BE DEAD. IT WAS A COOL, SUNNY MORNING WHEN we came out at 8 A.M. from last night's shift. I turned up the radio very loud, rolled all the windows down, and headed home. I awoke once when the car went off the road, tires crunching on the ashes. I thought then that I should have pulled over and slept, but it was only another mile or so to the trailer. I kept going. Suddenly I opened my eyes and I was on the wrong side of the road. A new green car was directly in front of me, sun glinting off the chrome, horn blaring. I didn't have time to do anything but heave the wheel to the right. The green car careened off the other side. We missed each other by inches. The other car didn't stop, just swerved back on the road.

Last night, Taylor, the pale, blond kid, and I were sent to clean out the runway along the main belt. Taylor complained that he was sick. He lay down to sleep on a pile of sand beside the big end roller, about six feet wide and five feet around. The sand was to be thrown into the roller if the belt started slipping.

There was no air at that end of the belt, since no one worked there. It was hot and stifling. The slightest exertion produced a sweat bath. I shoveled some of the muck onto the belt and sat down on an oil can. I was half dozing when I saw a flash of light at the door. Meroni, the boss, was coming. I kicked Taylor and he came up like a clay pigeon tripped out of a skeet toss. We were struggling with a piece of belt, trying to pull it from under the muck, when Meroni arrived. He was kind of smiling, probably because he had seen our lights dead against the roof when he came in. He didn't say anything, just tapped the roof with his hammer and told us to be careful. Later, Taylor went out, claiming he was sick. Meroni was enraged because I couldn't stay at that site without a buddy, and he didn't have any place else to put me. Finally, he gave me a sledge hammer and I went along the main shaft breaking rocks and throwing them against the side.

May 29

I MENTIONED TO A FELLOW IN THE LEGION THAT I HAD A near miss driving home yesterday. He worked in Greenridge and had to drive past mine #37 every morning, where my neighbor Larry works. He said he was always extra careful on that stretch of road because night shift was leaving as he came by. "It's like playing dodge car every morning," he said.

May 30

LAST NIGHT I WORKED WITH CLIVE SETTING KING RAILS IN the main shaft around the curve by the slope. As usual, we spent the first hour riding through the shafts on a motor looking for one of his tool stashes.

The shaft where we set the rails was about eight feet high. First, we dug down to solid bottom, about two feet, and measured the height of the roof. Then we sawed a prop. Next, since these rails were about fifteen feet long and weighed over a half ton each, we waited for Big Chuck, the motorman, to come by with a train of supplies. When he stopped, he and I lifted the rail onto the motor. We had nailed a piece of board across two props close to the roof, and we slid one end of the rail over the plank. Then Chuck put his back under the other end and lifted while I guided the rail up. Clive put the jack underneath and jacked the rail tight against the roof. Props were set under the rail and the board was hammered off. Then we moved to the next one.

The props were often measured too short or too long. Then the rail wouldn't have enough clearance to slide between the board and the roof, so we'd grunt and strain to lift it higher, only to find that it jammed against a rock hanging from the roof. Clive, wheezing mightily, would swing at the rock with a pick while Chuck and I groaned under the rail. If the prop was too short, we waited until Clive found a footer, a piece of plank, to put in the hole. If the prop was too long, we waited while he sawed an inch off. When a rail was finally up, there was jubilation.

June 3

INSIDE B-14, JUST OFF THE MAIN SHAFT, A LARGE TRASH pump in a dam about fifty feet long and thirty wide was working twenty-four hours a day to keep water from flooding out onto the tracks in the main shaft. Water flows into the dam from the abandoned long wall up at B-14. Coal continually washes down into the dam. If the coal gets high enough to clog the strainer, the pump shuts off and the main shaft floods. A small mountain of coal that had been shoveled away from the strainer was piled like a pyramid in the middle of the dam.

The roof was so bad that rails had been set every two feet. Vince told me to take a rail car and drive it with a motor as far as it would go without derailing into the water at the far end of the dam. Tracks ran under the muck and water all the way across the dam. Then I was to shovel coal away from the strainer and remove the pyramid of coal at the other end of the dam by shoveling it into a bucket, carrying the bucket across the dam, and dumping it into the rail car. Vince gave me a pair of waders, chest high, and a shiny new bucket. I laughed. I couldn't believe he was serious until he got mad about it.

I put on the waders, drove a rail car into the muck at the far end, and went into the dam. The water was freezing. It was like stepping into quicksand. The muck was knee deep and the water was just over my hips. I had to crouch under the rails. The roof was about five feet high, with a three-foot clearance between the water and the rails. With each step the muck gripped my legs. I held onto the overhead rails for balance. Each time I tried to pull one foot out, the effort drove the other leg in deeper.

I filled the bucket, two shovelfuls, and carried it fifty feet across the dam to the car. I was crouched over the water, one hand on the rails and the other holding the bucket, which floated on the water. Several times I almost fell. The viselike grip of the muck around my legs made slogging through the water very slow and dangerous. I pictured Vince returning at the end of the shift to check on me and seeing only a helmet and light turning lazily on the water.

Vince had given me a short-handled shovel, the only one he could find. To shovel the coal away from the strainer, I had to immerse my arms to the shoulder in the cold, fetid water. I would take three shovelfuls, then clap my hands and beat my arms against my sides to restore the circulation. The water ran down inside the waders and in no time I was soaked. It was a long, cold shift.

June 4

THE INJURY LIST TODAY: REMICK WAS OUT; A DISK IN HIS back slipped while he was lifting a rail. Billy Sweet was out with seven stitches in a finger. Aldo had smashed a finger with a hammer while hanging canvas; his finger was splinted and bandaged. Another man was out because a rock had fallen on his foot.

Meanwhile, I went back to the island of coal in the dam of icy, brackish water. The waders leaked. The left leg filled with water. I refused to use the short-handled shovel. Vince finally came up with a long-handled one. The catch was that the blades had all been straightened in the shop for shoveling under the belt. I still had to duck my arms into the water to clear the strainer.

I discarded the idea of carrying the coal across the dam by the bucketload. Vince agreed that it was a pretty stupid idea. I asked him whose idea it was to give me this job. He wouldn't look at me. Finally he mumbled, "Well, Cooper makes up the list." Suspicions confirmed.

At the end of the shift, I was in the shanty taking off the waders and putting on my boots when Helen and Joe Morgan came in. They had been in B-12, another slop hole, moving in cinder blocks. Helen was soaked. Her face was pale and lines of pain were sketched around her eyes.

"I'm hurtin'," she said. "I was throwing the blocks to Joe and when I threw one I hit my back against a rail. I can't lift my arm now. Doesn't this damn thing work?"

She was leaning against the radiator the mechanics had made, a row of welded steel strips plugged into a heavy fuse box. I threw the switch for her. "I'm going to the hospital for a shot of cortisone before I go home," she said.

She patted the radiator, coaxing it to warm up. "Cooper called me into the office today and said if I miss any more work they'll have to let me go. I told the sonofabitch I have a cold all the time and to keep me out of the water for a week. He looked at me like I was crazy."

Her voice was weak and scratchy and her nose was red. Morgan, his face sweaty and streaked with coal, sat on a toolbox digging dirt out of his lamp. He had been threatening to go to the committeeman if they did not take Helen away and give him back his regular buddy, Billy Sweet. He hates to stir up trouble, though. He told me while we were waiting for the cage that his latest strategy was to work Helen's ass off until she quit.

"Can you imagine one of us bitchin' about a hurt shoulder?" he said, his voice dripping contempt. "Christ, you work in here, you're always hurt."

Later, as we moved forward, I was standing beside

Helen in the line. She said, "You know what worried me most about coming to work in the mines? The other women."

I said that there was only one other woman in the place.

"Yeah, but we were told that there would be eight. Not that I really care about a reputation, normally I just say fuck 'em, but when it gets close to home, around my kids, then I care. Like, some of the other women miners were bodaggles."

I asked what bodaggles were.

"Lesbians. Male dikes. You know, butches."

We talked about her kids and she said they were adopted.

"That's how I got the twins, too. They're seven. I was working through an adoption agency for a year, you know, and then I heard that this girl was giving up her babies. Twins, a boy and a girl. They were six months old, four pounds each. Stink, Christ, you wouldn't believe. They had diaper rash all over when I picked them up."

Pam, in line behind us, was complaining because the girls had no showers. They had to change in the lavatory off the first-aid room, try to dry themselves, then each of them had a twenty-five-mile ride home. There was a shower in there, but Kurtz wouldn't allow them to use it. Everything was geared to forcing them to quit.

June 6

Cooper has evidently decided that I'll be in the sump dam for the rest of my days. I told Meroni that if they really wanted the place cleaned out, they would bring in the scoop and do it in a couple of hours. The coal washes into the dam about twice as fast as I can shovel it out. I have almost filled a ten-ton car, and there is more muck under the water than when I began.

When Meroni crawled across the planks that formed a bridge over the dam by the pump, I showed him that I had to put my arms under the water to get the shovel under the strainer which cleared the pump. Since I always had to crouch, with no chance to stand straight or sit, my back ached badly after an hour or so.

Meroni sighed and said, "Listen, I come in here and you give me hell. I go up to see Billy Sweet and he gives me hell. Joe Morgan and Helen are threatening to go to the committeeman. Goddamit, I'm the one with the buglight in here. I'm supposed to be giving you people hell. This job has to be done and that's it."

The water was down about a foot. Now it was around the tops of my thighs in most places. Pumper told me in the mechanic's shanty that Cooper had taken him up to the source of the water, "to see if we can't figure out this problem." As usual, Pumper said, he told Cooper what to do and Cooper went back to the bosses' office to tell them what to do, claiming credit for figuring it out.

We talked while I was pulling on the waders. Pumper said that one of my neighbors, a long wall boss, had made $30,000 last year by spending almost all of his time in the mine. So far, Pumper has made $9,600 this year working seven days a week. I've made $4,160, but I usually take a day a week off. Young Donchus, the buggy runner on Remick's crew, said his dad, who bosses the crew, made $26,000 last year. Then he laughed and said, "He's going for his inspector's exam tomorrow. He says if he makes it, he's comin' in and shut this hole down."

Not likely, Pumper answered. New inspectors weren't assigned to the mine they had worked in.

In the middle of the sump hole, crouching under bad roof, my lamp shone off the scummy water washing down the empty snuff cans, the gummy hawkers, and the shits the men took up at the long wall. I went back to shoveling the muck out of the water. It was impossible to wear gloves. Pulling up a shovelful was like landing a forty-pound fish; the viscous crap hung together until it broke the water.

Joe Morgan was working with Helen and Red. At 5 P.M. they had run out of work to do, so they dropped Helen off with me and went up to B-12 to pull out rails.

Helen sat on the gangway across the end of the water and sang country songs and talked while I shoveled out the muck around the strainer. She said she had been out yesterday because some kids had broken into her house and stolen $300. I felt sorry for her, a thirty-six-year-old divorcee with seven kids. Morgan was adamant about getting rid of her, and told her so. I said it didn't look like she was cut out for the mines, and asked what else she had in mind.

"I don't have anything in mind anymore," she said with a rueful laugh. "I just try to make it through the day. That's all I can do. I don't have the faintest idea of what I'm doing. I'll stay in here until I get hurt bad enough to go into the hospital, and then I guess I can draw compensation."

Then she said, "If I could keep all those kids some other way, I wouldn't be in here."

"If there was some other way, a lot of us wouldn't be in here," I said.

After the shift, I was walking toward the shanty and met Clive hiding his tools. I asked him why they had called him into the office earlier. His eyes were almost closed. He just shook his head and said, "Bitchin'."

Hank and Joe Morgan were in the shanty when Helen and I entered. During the conversation, Helen said to Hank, "Listen, if you don't get out of here now, you'll be trapped for the rest of your life."

"Don't worry," said Hank. "I'm only nineteen. I intend to get around and see a few things before I'm twenty-one."

Helen shook her head sadly, like a mother, and said, "Where else is he going to make fourteen thousand a year?"

We talked about the one vocational high school in the area. It was agreed that the kids graduated from all the public schools fit for nothing.

I asked Hank about the sexual revolution. He said he thought it was running backwards. "I got laid more when I was a sophomore than I do now," he said.

June 9

DAY SHIFT. THE WADERS WERE GONE FROM THE MECHANIC'S shanty. I told Crisco I wasn't going into the dam without them. He called outside to Cooper, who threw another pair on the elevator. They were several sizes too small. One of the mechanics said that the committeeman had thrown the other waders out because they were soft-toed. All soft toes are illegal underground. Those that Cooper sent down were also soft-toed, but I wore them because I didn't want to test Cooper's ingenuity in finding some far-off hole to stick me in.

The water was back up around my hips again. The

black scum sucked at my legs. Hal and Taylor crawled through on the planks, headed up into the B-14 shaft to drag belt structure out. It was a more dangerous job than mine; the roof was terrible and the wind up there was strong. Hal was hurt a few days ago when the spinner knob on the kersey's steering wheel spun around and cracked him on the arm. The doctor told him to take a week off and let it heal, but we are working now on the June 28 two-week vacation pay. He wouldn't receive compensation by that time, so he was working with a bad arm.

Later, I was at the end of the rail car shoveling muck from under the water. The doors at the other end of the car opened and somebody shined his light in. I thought it was one of the long wall bosses.

"Where's your bucket?" he said. "Scoop this stuff with a bucket."

"Nah, that's a stupid idea. It's easier with a shovel," I said.

Then I saw that it was Cooper. "Well, goddamit, get started then," he snapped, and went back out through the doors.

June 10

DANNY KURTZ DROPPED BY THE TRAILER LAST EVENING. He had been visiting another friend in the trailer camp. He and three girls weaved out of his van from which rock music was blasting. Danny wore big, square, pale-blue Elton John glasses. I admired his C.B. radio, an impressive piece of equipment. He said it was a nice thing for miners to have because when it rained they could sit in the parking lot and talk to each other without rolling their windows down. It was true; I've seen it. He and the girls came inside. One of the girls wore braces. Dan said he had been partying all last night, had gone to work today, and was going to another party tonight. I considered the advantages of being eighteen.

Grada had been at bridge club during their visit. We talked about Danny and his friends when she got home. She had never met Danny, but she knew him through our talks about the mine. Grada considers most modern teen-agers as lost souls because her own adolescence was a happy, active time with close friends, chatting about homework over the phone, ice skating parties down at the dam in winter, kind of a Currier and Ives childhood. It seemed almost a turn-of-the-century time, and she is only thirty-one now.

June 11

THIS MORNING CRISCO GAVE ME A HARD LOOK IN THE HALL and said, "Get back into the sump and get it done. You been lollygagging."

The men around us laughed. Everyone who goes across the plank bridge on the way up the shaft asks why I don't go to the committeeman and get transferred out of that place. Actually, having spent so much time alone with the island of coal, the cracked roof, damascene walls, and scummy water, I kind of feel attached to the place. My own dungeon in the middle of the earth.

But I nursed my indignation all through the shift, and at the end I met the committeeman in the shanty and told him I was wearing soft-toed waders. He whipped out his pad and pen, and, for better or worse, I knew it was the last I would see of the sump. One of the mechanics said, "Cooper will get on a man and ride him 'til the man quits. I've seen him do it. Look at George the timberman."

June 12

In the waiting room today, Crisco came over and said, "You're back in the sump."

"Not without steel-toed waders," I answered.

He sent me to Cooper, who was leaning back in his chair, hands behind his head. Well, said Cooper, there were some around somewhere. But I knew from Pumper that there were no steel-toed waders in the entire mine. Cooper looked in the lampman's shop, the super's office, and the locker where the rubber suits were kept. Finally, in his lazy, high voice, he said, "You go with Hal up to B-14."

Of course, he knew there were no steel-toed waders. He had used the search time to think of the right spot for me.

Hal and I walked up about a half mile to the structure. The roof, drummy and pouring water, was about a yard high. We knocked a hole in a brattice to push the pipe through into the parallel shaft where a kersey could take it out. Hal's swollen arm pained him. Pins and hooks hanging from the roof gouged chunks of flesh from our backs.

Finally we pulled out a couple of crib blocks, turned them dry side up, and broke for lunch. "This is what hell must be like," Hal said. "My wife keeps nagging me. I can't get it up for her anymore. This goddam place, like all the walking we did in the low to get up here, it kills me. Before I came in here, we could screw all night and then I could fish all day."

He talked about wanting to get out of the mines, and we discussed the school system which provided a constant supply of young miners.

"That's my story for sure," he said. "All I learned in shop class was how to make a gun rack. A fuckin' gun rack, and I could've been learning how to be a mechanic or something. And why study civics and chemistry? I still don't know shit about that stuff."

June 13

TWO PUMPS, ONE ABOUT 350 POUNDS, THE OTHER ABOUT 150 pounds, up in the B-12 shaft had to be dragged three hundred feet to the kersey. I went to B-12 with Joe Morgan. Helen, his usual buddy over the past weeks, was sent to the sump to shovel my island of coal. No one had told her that soft-toed waders were illegal.

Morgan is a quiet, affable man, intent only on doing his job better than anyone else. His perplexity on having Billy Sweet, his regular buddy who had been awarded the supply job after bidding on it, pulled out and replaced with a woman had turned into stolid discontent. He was aware that the women were shifted to different men until one was found who wouldn't storm into the office and threaten an immediate grievance, but Morgan just isn't that type. He had decided to work Helen so hard that she'd quit, yet he couldn't force himself not to do her work for her. He was glad we were going up to a low, wet slop hole to share a day's grunting and sweating. I was glad, too, to have someone to work with instead of being marooned alone in the sump.

Driving up, we had to crouch lower than the top of the machine or risk being thrown against the roof when the kersey hit holes and ditches in the low places. The crashes echoed down the shaft, lit brilliantly by the kersey's bright headlight. The shaft, oval-shaped like a mail slot, dipped and twisted, following the coal seam. It was a long ride, progressively colder, until we reached the pumps. We crawled through a brattice and pulled a two-wheeled dolly, the kind beer distributors use, back through the slop. The roof was between three and four feet high. After we put the big pump on the dolly, Morgan got in front and I pushed from the back. It was impossibly frustrating to wrestle with that much weight under low roof and in deep muck. Once Morgan took a shot in the kidneys from a low rail and dropped the pump. Later, I slipped against an eight-inch water pipe. The joint shifted and water gushed against the roof. We tied a rag around the joint.

When both pumps had been removed, we were soaked and exhausted. Proud, though. It took superior men to endure eight hours of that work. We either sweated or froze during the entire shift.

We stopped a few times on the way back to avoid coming out early. I told Joe what Hal had said about not being able to get it up for his wife very often.

"I got the same problem," he said. "That's one reason I got to quit working with Helen. My old lady always says I must be gettin' it from the women in here, because I sure don't give her any. When she gets bitchy and starts a fight, then I always know she wants it.

"You see these women around town, though, that's a

different story." He laughed. "My daughter, the oldest one, she's fourteen, and when her friends come up to the house, it shakes me up sometimes. By God, they didn't look like that when we were that age. All fuckin' now, too, they say. Maybe it makes them better lookin'."

I mentioned that Hal's father had told him of a doctor's opinion that walking in the low ruined a man's sex life. Morgan agreed.

When we went out, I stopped by the dam to see how Helen had made out. She wasn't there. It turned out that she had fallen into the water and gone home after two hours. Her parting words, from the mechanic: "Tell Cooper to go fuck himself."

Pumper always comes out early and hangs around the shanty. We were talking about accidents, and he said he was almost killed while wearing a dust pump, a small gadget with a purring motor and a long hose that clips to the clothes and measures dust in the air. Every miner is supposed to wear a dust pump once every three months. Most men take them off and hang them somewhere around the work site.

Pumper said he had been riding out at the end of the shift when part of the pump dropped over the side of the belt. He turned to pull it up and hit his head on a crossbar.

"It rang my bell, I'll tell you," he said. "If I'd of been knocked out, I would've gone right on out to the tipple. I told them that day, don't you ever give me another dust pump. I won't wear it."

June 18

No work until further notice. The company claims the cage is broken again. Some men say it's another layoff. I went down to Barley's bar in Cokeville last evening. Tom, the bartender at the union hall, and his older brother, Jim, were arguing with another aging miner about the old days. They were all drunk, laughing and kidding each other about how many mine ponies had been used at Crocker's mine. The old guy claimed that they had five hundred mules. Tom roared with laughter and elbowed me. "They must've had a big barn," he said, winking. When I was a small boy we used to ride an old mine pony around the pasture over the hill from my house. The ponies were used to pull carloads of coal from inside the mine out to the tipple. They were small and docile, and often blind from spending most of their time in the dark.

Jim maintained that they had been paid thirteen cents a ton in 1932 for pick coal and forty-five cents a ton for machine-cut coal. Somehow, they were able to carry the argument on all evening.

Dolly, Ann, and Barbara were at the other end of the bar. Dolly told me that her dream was to wait until her daughter was old enough to leave home. Then Dolly would go to work on herself, become beautiful, and marry rich.

June 25

We have been out a week because of the cage, and in another few days the annual two-week miners' vacation begins. According to Grada, who keeps our accounts, we have saved no money since I began underground, but we have managed to pay off all of our debts except for a new car.

We spent the day cleaning cupboards, scrubbing floors, polishing the windows and tables and the enamel stove and refrigerator. The louvered windows were cranked completely open, allowing a warm breeze to flow through the trailer. Grada, in one of my old T-shirts and paint-stained jeans, was scraping about ten years' accumulation of gunk from under the kitchen shelves. She started talking casually about how much we had accomplished in the past year.

"Honey," she said, "do you know that a third of today's marriages fall apart within five years? But we're still together and we're a terrific family. I mean it."

I agreed.

"We've accomplished what we set out to do, and in a couple of weeks we could save your whole paycheck," she continued. "If we weren't living here in the trailer, I mean."

We were on our knees at opposite ends of the open cupboards. She stopped scraping at a corner and carefully lined up a roll of shelf paper with the edge of the wood.

"I've put up with a lot this past year," she said. "You have, too, of course. But you don't want to stay in the mines forever, do you?"

"Nope."

"I was thinking that if I took the kids back to Jersey with me and stayed at my parents' until I could find an apartment, then you could live rent free out at the cabin with Baron. In a couple of weeks, by the time I found us a place to live, you'd have a couple of weeks' salary coming."

"Then what? What makes you think I'm more employable now than when we came back from Spain?"

She stopped working and set the roll of paper down. "Because you're different now. You're solider. And you're already doing the hardest work in the world. You could do anything you want. Have you thought about that?"

We took a break and sat at the kitchen table while she let me in on her plan. Her parents had told her that opportunities were scarce in the New Jersey–New York area, but they would be willing to help us until I landed something. Finally, if I couldn't find a job I could stay at home and Grada could look for something. After all, she had a degree in social work.

"I'm glad we had this experience," Grada said, "but enough is enough. We may not know exactly where we're going from here, but we have to make the move. There has to be a first step. You can't see anything from here, Meade. Moving out is a choice. It will make things happen."

We talked for the remainder of the afternoon. Grada had already planned the details of the move, and she intended to leave in three days. It became clear that she was telling me, tactfully, that she had had it and was leaving no matter what I said.

In the evening, I went over what I might do after quitting the mines and decided that newspaper and magazine reporting seemed like the best bet. Grada agreed, and we cleaned the suitcases and trunks for packing.

July 2

I AM ALONE AT THE CABIN. THERE IS NOTHING TO DO except take long walks through the woods with the dog, sit on the porch and read. This morning in the post office I ran across Ken Hodenart, a neighbor and father of a childhood friend. Ken is retired with black lung now, still living with his wife in an immaculate white house down the street from where I grew up. I was interested in the things he had to say about the old days in the mines, so I went back to visit and listen to him reminisce further:

"The Coal and Iron Police lasted from '27 to '28, until the last of '29. They seen to it that there weren't any pickets around. If you tried to congregate around the property, why, they rode you down with a horse and gave you a smack on the head with their billies.

"It was forty-seven cents a ton for pick coal back in '32. The thing was at that time they were able to handle only so many cars that came into the mine, and they divided them among the men. The little wooden cars held a ton three, the long ton. But they were a ton two, three, or four. But you never got a ton for one of them. The company demanded several inches above the car. They short-weighted you, you better believe it. Just before the union came in, I was working in a little mine out by Slag Hollow. The coal that came out, that you could see in the cars, you could figure it out. In this one week, they loaded four or five cars, fifty-ton cars, and the weight was between forty-five and fifty tons in those cars. With all those cars, they paid those men off for just over a hundred tons.

"Where I worked, right up against the roof there was a streak of boney, then atop that another streak, a rider of coal. Atop the boney they called it top coal. But you had to separate this top coal from the other coal, and load it separate. Pile it along the rib. The machine came in and cut the place, and you piled the cuttings along the rib. Then one day they'd say, okay, today we'll accept the cuttings, and you'd load it. Or today you can load your top coal. My dad and I loaded these cuttings and top coal, eight cars in this one day, maybe a ton to a car. But we loaded them cars and did the rest of our work. I came out the next morning and they didn't mark it on the check sheet. Well, I thought something happened, they didn't mark but one car. I asked them what happened to the other seven cars, maybe they switched them out or something. The next day I looked again, still they weren't there. I went in to see the guy and said we sent out eight of these cars, cuttings and top coal, and so far we only got one marked. He said, oh well, they cut the price down on it, I could pay you for only one. What do you do? Smack him in the mouth and lose the job, or accept it. So, we accepted it. It was something they stole off me just in that one instance. There was just no recourse. None. The same man we worked for, same company, they took us into another place, a new area, to make a heading.

"There was eighteen inches of rock to shoot. He said he'd give us ninety cents a yard to shoot this rock and throw it aside, or a dollar fifty a day company wages. Well, my dad was familiar with this work, he did a lot of it, and he said we'll make out better at ninety cents a yard. We'll take that rate. The owner said he would furnish the powder. So, we shot the rock off, loaded it aside by hand, and we did the job in two weeks. We used a force auger. So the three of us did this. I was twenty-four, my brother was twenty-three, and my dad, and the day they came around, we were done. He said, I better measure this. He took a tape out of his pocket and he said, 'Oh boy, you guys made all kinds of money, but I can't pay you for this. I'll have to pay you just straight company wages.' So, there you were. We made too much money at ninety cents a yard, so he took that off us and paid us straight company wages, a dollar fifty a day. Two hundred feet, eighteen inches thick, fifteen feet wide, ninety cents a yard was too much for that.

"As far as I can see, the attitude is still to use the man up and throw him away. Same attitude. Now the federal is in it, the state is in it, but it's still the same attitude as far as I can see. Now, the judges, the lawyers, even the doctors, people who have never seen a lump of coal, they're getting rich off shunting black lung people around. Every time you make a move, if they don't make money off you, they make it off the union. It's tax money. Somebody along the road is paying them, insurance, whatever. So, if there's concern about whether you're eligible for black lung, each time you make the rounds somebody is paying these people. Even when I was working, for a regular compensation case, if you got a finger hurt and went to the doctor, that was twenty-five dollars.

"For the first visit. So you can see what they're making on it. When I got my black lung, the attorney I went to, he sent me to a doctor, the state sent me to one, the company sent me to one, that's three, then there was a neutral one, that's four, and then the federal sent me to one. I just got a black lung check today, $111.43, every two weeks. This plus twenty-one dollars from the federal for black lung. I get social security, plus miners' pension. The only thing is, it was three years before I got the miners' pension. I was bossing for a while, and my records were kept so piss poor by the company, I was held up three years. They didn't separate my time as a fire boss, face boss, whatever. Then the owner died.

"If you went into a place for dead work, you didn't get paid for that. If you got into an area where the bottom was heaving up, well, if you were lucky and followed a shift in that took the bottom up, you might get away without doing it, but you could figure out of every day one shift put in two or three hours, maybe the next shift nothing, the third shift another two or three hours. It would heave up, every minute it was coming up. You had to take it up to get your car through. You did enough to get your car in and out for that day. You had to move these cars by hand. The most they run in was three hundred feet. You just kept pushing this car.

"If it didn't go, then you stopped and brought it back and took up some bottom, then took it up again. Sometimes you'd be working and you'd hear a big bang, the earth was heaving up, and you'd know you'd have to quit early to take bottom up so you could get your car out. It sounded like a half case of dynamite going off in there. You sat there for a while until the dust cleared, you couldn't see nothing for a while. Now they have canvas in the crosscuts, they direct the air. Then they didn't. The only air movement was you pushing the car up there.

"Whenever you shot, you stayed out for a while. On your backbone, every vertebra was blue from pushing the car, you'd dig your heels in and push with your back. If the push was too hard, then you couldn't get your car to the face. So then what the hell did you do? You sat down, you put your back up against it, and you started pushing that way. Up a couple inches, then your buddy pushed a couple inches. You took turns.

"I worked a seam one time where I put my foot on the bottom and my other foot on top of it, and my toes just brushed the roof. Of course, we weren't there too long. The boss had to find out from outside, if the men can't do nothing with it, you moved. If it was an important area, they'd pay you dead time to get through it."

July 15

IT IS 5:45 A.M. THE FIRST DIM, GRAY LIGHT APPEARS IN the sky. Rain patters softly against the cabin roof. White mist hangs over the creek. Crows call in the distance. The dog sits at the screen door, listening, cocking his head. The cabin smells of fresh coffee and dampness. I have been here alone for two weeks, and have done nothing. Now it is time to leave for the first day back at 18-D.

By the time I pulled into the parking lot, patches of blue showed in the sky, and shafts of pale, early sunlight were breaking through.

Inside, everyone was glum. Hank, Clive, and I were sent to Main A to replace rotted props. Clive went farther down by himself because three men just get into each other's way.

We set only ten props and quit. We walked down to where Clive was setting props in a water hole which we had passed up earlier. He was mumbling about the hole for the prop filling with mud and water before he could get the new prop into it. Hank said that the cage actually had been broken, and his uncle had been ready to send Danny and him to Chicago to pick up the parts. Then they decided to fly in the parts. We helped Clive set the prop. After a month off, it was hard, working with thick, seven-foot props, digging holes in the slippery, gummy mass, scooping out the holes by hand, smelling the stink rise from them. We had new jacks now, unreliable screw types instead of the hydraulic pump jacks. It is impossible to tell when the screw types are tight enough to keep a rail from slipping out when the prop is removed. But cheap is the magic word in 18-D. They are also using a new rock dust, gray instead of white, five percent free silicate instead of one percent in the previous limestone.

Hank hit the rib, the side, with his hammer to test it and a 150-pound chunk fell off at his feet. He kept regretting that he hadn't gone to the Pillartown steel mills during the layoff. They had called him about an application he had made. He said he didn't go because they laid off too much. Then we tried to add up the time we have been off this year. Although we couldn't pin it down, the steel mills couldn't possibly have done worse.

Hank told me what had happened to the cage. Crisco had been at the bottom. He pressed the button to bring the cage down. It came down, a cable snapped, and it kept right on going, smashing on the springs at the bottom of the shaft. That shift was called out, and the rest of us had been told by phone not to come in.

After we had set the prop, the three of us walked down to wait for the mantrip. The rest of the crews were already there, sitting on their buckets and telling stories. It seemed that everybody was quitting early the first few days back.

When the mantrip came, everybody piled in. On the way to the cage, we were flagged off up ahead. There was shouting and cursing up front. Word came back that the cage was down again, and we had to walk out the slope.

It was warm outside, but the stiff wind still turned the long, forty-five-degree walk into an endurance test. I walked the entire slope without stopping. When I emerged, dizzy and with a pain in my chest, into the bright sunlight and muggy eighty-five-degree heat, crowds had formed to wait for the trucks to take us to the washhouse. The most aggressive men jostled for position. A dump truck swung into the lot and stopped. A crowd ran at it, elbowing each other, shoving, climbing into the bed, filling the truck in a half minute. Several men had to be thrown off before the driver would consent to leave. When it pulled out, Aldo was dead center, smiling and waving.

Hank motioned for me to look behind us. Clive had just emerged from the slope and was leaning over a car fender, gasping, eyes closed. He was almost bent double, his face contorted with the pain of breathing.

I caught the next truck. In the washhouse, Cooper told us that the cage would be down for at least a week.

July 19

NOTHING TO DO TODAY. I WENT INTO TOWN TONIGHT, Saturday night, for the first time since we were laid off four days ago. The bars were crowded. I met Hippie Joe and accompanied him to a Sons of Italy dance in Tippleside. I once played that place fifteen years ago. Nothing has changed. When you ask a girl to dance there, you're stepping out onto the floor with somebody's daughter, and everybody in the room knows it.

Downstairs in the cellar, the men played murra; one, two, three: and hands jab across the table, fingers trembling, the men shouting *sei!* or whatever. They took it very seriously.

Hippie Joe said that last night some guys from the

Legion had taken off for their usual run to the Springer Hotel and just kept going. They turned around at Washington, D.C., and didn't get back until 10 A.M. this morning. Several of the married men won't be let out for a while after that caper, he said.

July 22

We went back to work last night on hoot owl. I had my old easy job of picking out the dirt from the tracks in the main shaft.

In the washhouse, they were talking about what had happened when the cage broke. Jack, an old, bald fellow, had been riding down on it. The cage didn't stop, and he went slowly to the bottom of the shaft. He called outside on the elevator phone and asked what to do. They instructed him how to bring it up, and bring it up he did. It kept going to the top of the shaft and smashed into the roof. They said that the cage moves from six hundred to nine hundred feet a minute. He was thrown against the top of the cage and hurt his neck. Then he had about two feet of clearance to crawl out. They said he came out, picked up his bucket, and went home without a word to anyone.

July 24

No work, cage broken again. My car radiator leaked, rotted out, so I went to a junk yard this afternoon on a hill covered with rusted hulks. The sun glinted off cracked windshields among waist-high weeds, overgrown scrub, and ragweed. Chickens, cows, and two turkeys milled around the owner's house at the bottom. I walked over the hill for a couple of hours, kicking at the weeds to chase out the snakes, then came across an upside-down Ford with a child's doll wedged under a crushed seat covered with bloodstains. The whole hillside was a monument to violent death. The radiator cost $10.

July 29

WE WENT BACK TO WORK ON SECOND SHIFT, 3 P.M. TO 11 P.M. On the bench in the waiting room, Helen said she'd like to talk to me after work. I said okay, we'll have a beer at the River Gap Hotel.

Cooper sent me down along the side of the main shaft to sledge rock. It was an easy job; I could stand up. I ate lunch with Billy Sweet and Pam, who were shoveling gravel from a rail car onto the passage to improve footing. Billy and Pam have been buddies for about a month. It is remarkable how it has improved Billy's disposition. He comes out whistling on some days, and he couldn't care less about the gossip. She treats him as though they have been married for thirty years: scolds him, teases him, exchanges stories with him. Pam said that Helen wanted to tell me that she had been called into the office by Cooper and Kurtz and told that there had been complaints about her work. Helen thought that Pam, Billy, and Joe Morgan had been doing the complaining, Pam said.

After work, I stopped at the hotel. It was almost midnight, warm and breezy. The bar was dark, lit only by the color TV and the Budweiser clock over the kitchen door. I bought two slices of pizza and a draft.

Helen arrived about twenty minutes later. She sat on the stool next to mine and ordered a draft. She was nicely made up with blue eye shadow, and her black hair was brushed and shining. I asked her what had gone on in the office. She said Kurtz had been on one side and Cooper on the other, badgering her about complaints they said they had been getting.

"I talked to the men I worked with," she said, reeling off several names, "and I got along good with every one of them. It's Pam, sits around with Cooper all the time, who's behind all this. Her and Billy want to work together. I'll tell you, she's been through a lot of men already at this mine. One of these wives is going to get her yet. But she's the one behind it all. I mean, I'm not a psychologist or anything, but she wants to be the only woman at this mine. She wants to be the center of attention, that's why she tells those really dirty jokes. I know she and Billy told Cooper and Kurtz that they wouldn't work with me anymore, that they didn't want to work with me. I figure I have to stay in this mine a year and a half to get what I want out of it and get my life straightened out, but Christ, it's a fight all the way. I don't know what to do. After Cooper and Kurtz told me all this shit, I went down to work with Joe Morgan. I was going to tell him about it, but my eyes kept filling with tears. I kept my back to him on the motor until I could finally ask him if he had gone into the office and complained about me. He said no, absolutely not."

Her voice became husky and she looked at the back of the bar instead of at me. I offered a piece of pizza. She refused and continued.

"I mean, I know Morgan's point. He says now that

when he doesn't want to go shopping, his wife right away says, 'Oh, you won't go shopping with me, but you'll go to work underground with another woman all day,' and he has to take all this shit from the men about working with me. But Jesus, what am I supposed to do?"

I said the first thing she should never do is go into the office without a committeeman along.

She said she had asked the committeeman for a copy of our contract three times but he wouldn't give it to her. "I told him if you don't want to represent me, give me my dues back."

When she was finished, I asked her if she minded if I told Billy and Pam what she had told me, just to get their reaction. She said no, she didn't mind.

August 5

BILLY AND PAM DENIED EVERYTHING. PAM HAS BEEN removed as Billy's buddy and sent to the long wall to replace Danny Kurtz. Danny was taken off the long wall again because he missed so much work.

I was put on Lou Perski's crew today as the pinner helper. At the beginning of the shift, we sent up fifteen planks, a half car of props, and thirty pins (roof bolts) on the belt. The crew was closely united in their hatred of Perski. Giving instructions, he exudes an arrogant contempt for all of them. In the mantrip going in, one of the young bosses was telling Perski about a time when another boss had saved his life during a cave-in. Perski said, "Well, the dumb bastard. We'd of been rid of you."

I helped the roof bolter. It was low roof, but dry. We finished one room and moved into another that had such bad roof that it shattered every time the drill entered it. We stayed behind the bolter to drill, but I had to go under the bad part to change steels, the drill parts, and to put the bit into the hole. Finally I refused to go under the roof again. The bolter agreed, and we spent the remaining hour of the shift avoiding Perski.

In the washhouse, after our showers, the long wall men were discussing how Pam was making out up there. She had not dragged a single dowdy jack all day, but the long wall helper, a gruff little guy nicknamed Baggy, was willing to give her the benefit of the doubt. "Maybe she was flaggin' today," he said, referring to her period.

On the way out, I glanced at the injury list. Remick will have an operation to remove a disk from his back. Endive will have a hernia operation. Glover, the new kid, sprained his back jumping off the belt when it didn't stop after he had pulled the jabco. Jack, the man who had been on the elevator, was out with a sprained neck.

I went into the office and found Kurtz and Cooper. I told Kurtz that I would be quitting on August 15th. He was at his desk and Cooper sat on a bench against the wall.

"My wife has already moved to her parents' place in Jersey," I said. "I'm going down there to join them."

Kurtz nodded, and in a frank, open manner said, "Okay, Arble. I wish you luck in whatever you do."

I nodded back and said thanks. When I turned to leave, I looked at Cooper and noticed that his face was very flushed. He was staring hard at me. I wondered what was running through his mind, and I had the notion to walk over and shake his hand. There was a pause while we exchanged looks, and I found that I couldn't get past the vibes he put out. It was too late for us to change conceptions now anyhow—we would forever think of each other as assholes.

I left the washhouse and walked across the red-dog parking lot in the warm afternoon sun. The car windows were up to keep out the dust, and the interior was baking. I rolled down the windows, pulled slowly out of the lot, and ran my fingers through my hair so it would dry on the way to the cabin.

I turned on the radio and the newscaster said there was a sniper loose in the area. In Fairview, near the Springer Hotel, he had killed an old man and wounded two other people. Police suspected him of using a Starlite Scope, an infrared scope used in Viet Nam for night sniping. Well, I thought, it's nice to have the dog out there. A big black German shepherd listening at the screen door was the best possible defense against intruders. I kept a .32 automatic pistol handy in the bedroom. For the rest of the ride to the cabin, my mind worked out a fantasy of hunting the sniper through the woods with dog and pistol. He wouldn't stand a chance; I knew every gully and nook on those eighty-two acres and beyond. Besides, if I didn't convince myself that I could win, I would be unable to stay out there.

August 15

MY LAST DAY WAS SPENT WITH THE CREW IN B-15, TRYING to run the Lee Norse up a steep grade into the face. The roof was terrible. Water ran everywhere and all of us wore wet suits which were quickly shredded by the low roof. The crew had been there for six months, and they expected those conditions to exist for another three.

That evening I met Zurko, Hippie Joe, and Paul in the Greenridge Legion. We drank beer and said our good-byes. Hippie Joe had wedged his stool into the corner of the bar and the wall. He was telling a long, rambling story about fishing. His voice was so graveled he sounded like Louis Armstrong. Meanwhile, the bar was slowly filling for the ten o'clock drawing. Behind us, next to the clanging pinball machine, an amicable argument erupted at the dart board.

I left early, before the drawing, and walked back past the old brick elementary school and my car. I crossed the street and continued down to the black asphalt strip that divided the cemetery and ended in a loop about halfway back. A warm breeze had arisen, turning over the dark leaves of the maples. Thin, translucent clouds scudded fast across the moon.

Off to the left, my father's small, granite headstone shone clean and new beside the weathered stone of his father. I walked to the two graves and stood there listening to the wind.

For a second I thought I could hear my father say, "I'm proud of you, son." He'd never said that to me in his lifetime. And I'd never really been proud of myself all these years. But now, as I stood there saying good-bye to the past, and to my ghosts, I did feel proud. Damned proud. During this year I had come to know myself and to understand other men. The future might bring the same problems that had defeated me before, but I felt confident. I had come through the long tunnel, and I'd found an exit.